RACE TO MARS

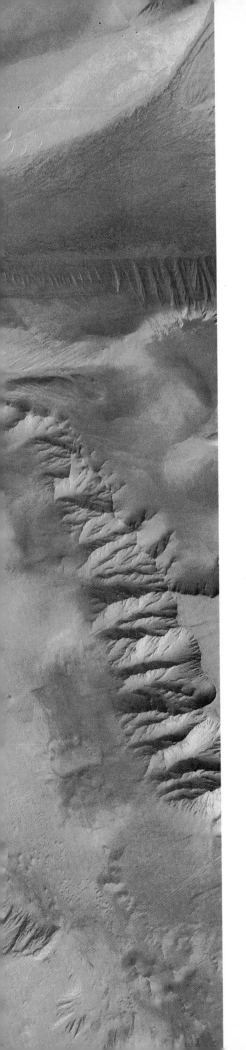

RACE TO MARS

THE ITN MARS FLIGHT ATLAS

General Editors
Frank Miles   Nicholas Booth

MACMILLAN
LONDON

Copyright © 1988 Roxby Productions Limited

Created by Roxby Productions Limited
a division of Roxby Press Limited
126 Victoria Rise
London SW4 0NW
In association with ITN Independent Television News Ltd
48 Wells Street
London W1P 4DE

General Editors: Frank Miles and Nicholas Booth
Text Editor: Chris Cooper
Design: Neil Clitheroe, Direct Input Ltd
Typesetting: Hobbs the Printers Limited, Southampton
Origination: J Film Process Co Ltd
Printed by New Interlitho in Italy

First published in the United Kingdom 1988 by
MACMILLAN LONDON LIMITED
4 Little Essex Street London WC2R 3LF
and Basingstoke

Associated companies in Auckland, Delhi, Dublin, Gaborone,
Hamburg, Harare, Hong Kong, Johannesburg, Kuala Lumpur,
Lagos, Manzini, Melbourne, Mexico City, Nairobi, New York,
Singapore and Tokyo

British Library Cataloguing-in-Publication Data
Race to Mars: ITN Mars flight atlas.
1. Mars (Planet)—Exploration
I. International News Limited
919.9'23   QB641

ISBN 0-333-46177-0

CONTENTS

PART·ONE
THE MISSION

INTRODUCTION

The race to land humans on Mars has begun. Both the United States and the Soviet Union have publicly announced their intention to carry out manned missions to the red planet.

The Soviet Union intends to land men on Mars by the end of the century. Their new Energia booster, the most powerful rocket in existence, gives them this capability. The Soviets have unparalleled experience of manned spaceflight and hold all the long duration records. The 1988 launch of two unmanned probes to investigate Phobos, the larger of the two Martian moons, will be followed by ever more ambitious ones. During the 1990s, the Soviets will survey the red planet in unprecedented detail, to prepare for manned landings, using unmanned roving vehicles and balloons, eventually returning samples of the Martian soil back to Earth.

The United States space programme has been faltering in the wake of the Challenger disaster in January 1986. NASA's standing and funding has declined since the spectacular success of the Apollo missions to the Moon. Yet, there have been repeated calls to re-invigorate NASA with a manned mission to Mars as its focus.

The historical parallel with the race to the Moon is striking. Within the first five years of the space age, the Soviets launched the first satellite, the first animal and the first human into space. In May 1961, spurred on by these spectacular events, President John F. Kennedy committed NASA to send men to the Moon by the end of the decade. In July 1969, two astronauts walked on the lunar surface for the first time.

Many options for sending men to Mars have already been proposed in the US, including quick 'sprint' missions and a highly elegant use of the geometry of the orbits of Earth and Mars to allow spaceships to repeatedly cycle between the two worlds. A growing awareness of the Soviet lead has urged presidential space policy advisers to direct NASA to develop the technology needed to go to Mars.

Given the magnitude of the undertaking, many people suggest that the first manned mission to Mars should be jointly undertaken by both superpowers. Cooperation in space is not new, but has been mainly restricted to unmanned research. The only manned space 'spectacular' to date was the linking of Apollo and Soyuz capsules in Earth orbit in July 1975. Heralded as a great breakthrough, it was suggested that both nations would develop common docking systems so they could rescue each other's astronauts from stranded spacecraft. But nothing came of it: no designer has so much as drawn a circle on a piece of paper.

Despite worthy claims for a joint manned mission to Mars, it is hardly likely. Space has always represented a high frontier, another arena in which the competing systems of government can demonstrate their superiority. It was for this reason there was a race to the Moon in the sixties, and why in the closing years of this century, Mars will be the focus for another. So while it is certain that the first astronauts to set foot on the red planet will carry the flag of one country only, it is an open question whether it will be the Stars & Stripes or the Hammer & Sickle.

The difficulties in going to Mars are not of the same order as those that faced the Apollo astronauts.

As we will show, travellers to Mars will face many unknowns and dangers. Unlike the few days taken to journey to the Moon, a mission to Mars will take around three years to accomplish.

There would be little point in sending astronauts there without the setting up of a permanent outpost. The race to Mars will signal a permanent human presence in space, so perhaps future historians will record the summer of 1988 as a milestone in the evolution of mankind. The exploration of Mars will see the first steps in colonizing the solar system. Apart from our own Moon, Mars is the only planet which could conceivably be colonized. Yet it is still a harsh, alien world requiring much expenditure of effort and energy to be made even remotely habitable.

To many people, such an enterprise may seem futile: why go to

Mars? Once again, history provides us with a telling precedent. The reasons are very similar to those of the sixteenth century explorers who set sail on journeys of comparable length and danger to open up the new world; a desire for national prestige, a sense of adventure, but most of all, for economic gain. As natural resources on Earth dwindle, we will look to other planets for the wealth of materials and energy sources they have to offer. Mars and its moons represent the first sites for mining and industrial development of our own 'new frontier'.

A manned mission to Mars will be the supreme technical achievement of a century dominated by technological innovation. The consequences for humanity could be as profound as the evolution of our earliest ancestors into *Homo sapiens*.

An eerie sunrise on Mars captured by the cameras on board the second US Viking lander—a sight that will soon be viewed by man.

The Soviet Union has been interested in going to Mars since the start of the space age, 30 years ago. The nation that had been the first to place a satellite in orbit around the Earth and the first to send a probe past the Moon was determined to be the first to the red planet.

As early as October 1960 two attempts were made to send probes to Mars, but the craft failed to reach parking orbit around the Earth. The Soviet Premier, Nikita Khrushchev, had timed his visit to the United Nations in New York to coincide with these attempts, and he is alleged to have taken a model of the spacecraft with him. It was never displayed.

The Soviets enjoyed greater success in other areas the following year. In April a Vostok rocket launched Yuri Gagarin on the first manned spaceflight, consisting of a single orbit. Two months before that, a version of the Vostok with a fourth stage had been used to send a probe to Venus.

Further attempts

Opportunities to launch probes towards Mars—known as 'launch windows'—arise at intervals of 25 months. Three more attempts were made during the next launch window, in 1962. The first and third of these craft reached parking orbit but, when they were fired for the Mars journey, failed to gain Earth escape velocity. These attempts were not announced by the USSR. The second craft, launched on November 1, headed off towards Mars and was designated Mars 1. Unfortunately, though it was later calculated to have passed Mars at a distance of 120,000 miles (193,000 kilometres) on 19 June, its transmitter failed 140 days after launch. Zond 2, launched during the next window 25 months later, suffered a similar fate.

Past Soviet Missions

Year	Mission	Launch date	Arrival date	Comments
1960	1960A	10 Oct	—	Failed to reach Earth parking orbit.
	1960B	14 Oct	—	Failed to reach Earth parking orbit.
1962	1962A	24 Oct	—	Failed to leave Earth orbit.
	Mars 1	1 Nov	June 1963	Flyby mission; contact lost before arrival.
	1962B	4 Nov	—	Flyby mission; failed to leave Earth orbit.
1964	Zond 2	30 Nov	Aug 1965	Flyby and possibly lander mission; contact lost in flight April 1965.
1965	Zond 3	18 July	—	Communication test to Mars distance; successful.
1969	1969A	27 Mar	—	Possibly lander mission; failed to reach Earth orbit.
	1969B	14 Apr	—	Possibly lander mission; failed to reach Earth orbit.
1971	Cosmos 419	10 May	—	Orbiter/lander mission; failed to leave Earth orbit.
	Mars 2	19 May	27 Nov	Orbiter successful; lander failed.
	Mars 3	28 May	2 Dec	Orbiter successful; lander functioned for 20 seconds on surface.
1973	Mars 4	21 July	10 Feb 1974	Orbiter mission, but flyby made instead because of retro-rocket failure.
	Mars 5	25 July	12 Feb 1974	Orbiter mission; successful.
	Mars 6	5 Aug	12 Mar 1974	Flyby/lander mission; contact with lander lost before reaching surface.
	Mars 7	9 Aug	9 Mar 1974	Flyby/lander mission; lander missed planet.

Pictures from space

But these communication problems were overcome in the following months. Although Mars was no longer in a favourable position, the Soviets were able to check this with Zond 3, launched on 18 July 1965. It took 25 pictures of the far side of the Moon and transmitted them from time to

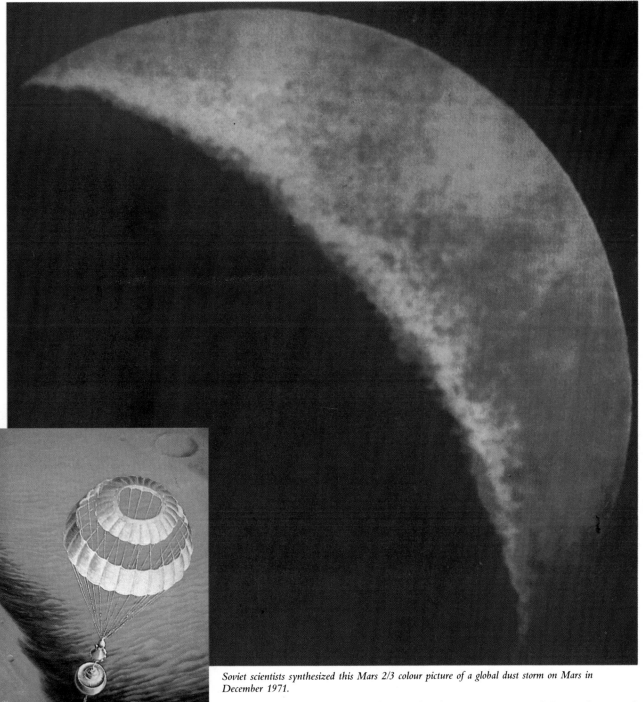

Soviet scientists synthesized this Mars 2/3 colour picture of a global dust storm on Mars in December 1971.

The Mars 3 lander descending by parachute before the retro-motors below the canopy fired for a soft landing.

time as it sped away from the Earth. In due course it crossed the orbit of Mars, and was still transmitting.

At the time of the Zond 3 launch the USSR was testing its more powerful Proton rocket. By 1969,

following a programme of mixed success, it was ready to try again for Mars. It is believed that two or perhaps three attempts were made to launch probes, though none succeeded in reaching orbit.

When the Mars launch window opened again in May 1971, the Soviets tried again. The first of three attempts again failed, the craft remaining stranded in parking orbit. It was given a coverall 'Cosmos' designation to hide its real purpose.

But the other two attempts of that year, Mars 2 and Mars 3, were

launched successfully and proved to be combined orbiter-landers. Mars 2 went into orbit around Mars on 27 November, followed by Mars 3 on 2 December. Quite different orbits were achieved; Mars 2 made one orbit of the planet every 18 hours 20 minutes, whereas Mars 3 took 11 Earth days. Photography was hampered by violent dust storms that were raging when the orbiters arrived, but both performed close to expectation, providing data on the temperature and composition of the Martian atmosphere.

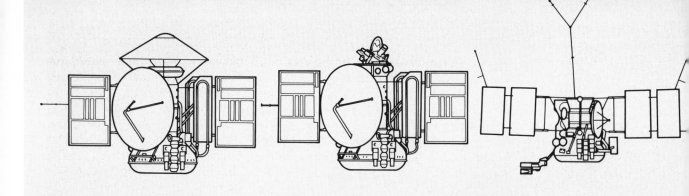

Mars 2/3, 6/7

*Purpose: 2/3: orbiter/lander
          6/7: flyby*
*Flown: 2/3: 1971–2
        6/7: 1973–4*
*Launch mass 2/3: 10,250lb (4650kg)
            6/7: 7720lb ? (3500kg ?)*
Height: 13.5ft (4.1m)
Span: 19.7ft (6m)
Solar panel area: 69ft² (6.5m²)
Lander diameter: 4ft (1.2m)
*Payload: landing capsule 990lb (450kg)
          surface/atmospheric cameras and
          scientific instruments (2/3 only)*

Mars 4/5

Purpose: orbiter
Flown: 1973–4
Launch mass: 9670lb? (4385kg ?)
Height: 13.5ft (4.1m)
Span: 19.7ft (6m)
Solar panel area: 69ft² (6.5m²)
*Payload: TV cameras and scientific
          instruments for imaging surface
          and probing atmosphere*

VeGa 1/2

*Purpose: Venus flyby, Venus lander and balloon,
          Halley's comet flyby*
Flown: 1984–6
Launch mass: 10,800lb (4900kg)
Height: 10ft (3m) without Venus landing capsule
Span: about 30ft (9m)
Solar panel area: 107ft² (10m²)
Data rate: 65kbits/sec
*Payload: Venus: lander for soil analysis
                 balloon to analyse atmosphere
                 and measure winds*
        *Halley's comet: TV cameras and
                 instruments for chemical and
                 dust analysis*

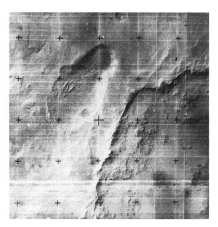

Above and below: low-resolution views of craters in the southern hemisphere of Mars, taken by Mars 5 in February 1974.

Above: Mars 2/3's landing capsules were hidden beneath conical heatshields during cruise. The tubes at bottom front contained cameras and other instruments.

Signals from the surface

The landers were not so successful. They were not reprogrammable, and the landings could not be deferred to avoid the dust storms. Contact with the Mars 2 capsule was lost just before it reached the surface. Signals from the Mars 3 capsule, relayed through its orbiter, ceased after only 20 seconds. It is assumed that it was overwhelmed by the dust. Nevertheless, these were the first manmade objects to land on Mars, and the achievement was the Soviet Union's.

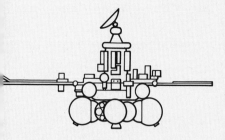

Phobos

Purpose: Mars orbiter and satellite lander
Flown: 1988–9
Launch mass: 13,200lb (6000kg)
Height: about 11.5ft (3.5m)
Height without propulsion module: about 8ft
*   (2.5m)*
Span: about 33ft (10m)
Solar panel area: about 107ft² (10m²)
Data rate: 4kbits/sec
Payload: TV cameras and scientific instruments
*   for imaging surface (Mars and Phobos)*
*   and probing atmosphere (Mars). Phobos*
*   soil analysis by two landers.*

The evolution of Soviet deep-space craft

Phobos is based on a deep-space probe design with an impressive record of success since the teething troubles of the early 1970s were overcome. Seven have been launched to Mars and 10 to Venus (including the two VeGa craft that flew on to Halley's comet). In addition, a modified Astron version successfully carried an astronomical telescope in Earth orbit in 1983. Since the failures of the 1971 and 1973 Mars missions, the spacecraft has enjoyed complete success on the Venus missions of 1975 to 1984.

The design is based on a central cylinder housing propellant tanks and a pressurized torus 7.5 feet (2.3 metres) in diameter and containing control and communications equipment. The main rocket engine fires through the central hole of the torus.

For Mars and Venus lander missions an encapsulated descent package was mounted on top. In place of this, the orbiter and flyby versions carried observation payloads, including the large mapping radar antennas of Veneras 15 and 16 in 1983. Later spacecraft had additional solar panels for increased electricity supply.

Although Phobos at first sight looks radically different, it comprises a structure of the same basic shape atop a large new propulsion module. The electronics, radio and control systems have been updated to create a new basic model for the 1990s.

Mars 3 was equipped with a parachute in a doughnut-shaped container, and the landing capsule was housed beneath a conical aeroshell used for braking in the Martian atmosphere. There were large solar power arrays and a high-gain dish antenna for receiving commands from Earth

The orbit of Mars is eccentric, with the result that the planet's distance varies greatly from one launch window to the next. By the time the next window opened in July 1973, the distance was so great that the Proton rocket was not capable of sending combined orbiter-landers. So two orbiters and two landers were launched separately, with designations Mars 4 to Mars 7, in a period of just under three weeks. All four shared the same basic design, similar to that of Mars 3 in having a large antenna and solar arrays; but each orbiter had imaging and remote-sensing equipment in place of a lander's aeroshell, capsule and parachute assembly.

The retro-rocket on Mars 4 failed to fire when it reached the

planet, but the probe took pictures as it flew past at a distance of 1370 miles (2200 kilometres) on 10 February 1974.

Mars 5 arrived two days later and was successfully placed into an orbit with a period of nearly 25 hours, a little more than one Martian day. This was the only one of the four spacecraft to accomplish all its mission objectives. As well as acting as an orbiting relay station it provided data on the composition of the atmosphere, and colour pictures from which maps of the surface were subsequently produced.

Mars 6 reached the planet on 12 March and provided the first direct measurements of the pressure, temperature and chemical composition of the atmosphere, but contact was lost 148 seconds after

When Mars 3 was safely on the surface, protective panels were to open so that pictures could be taken through the rotating turret on top.

the parachute opened, just before touchdown.

Mars 7 had arrived three days earlier, but although it successfully separated from its 'bus', a system malfunction caused it to miss the planet by 800 miles (1300 kilometres).

The Soviets decided to leave Mars alone until their space technology had improved. They turned their attention to Venus, which is closer to the Earth and hence requires a shorter journey time. They brought off a series of resounding successes with Veneras 9 to 16. These were followed by the two VeGa spacecraft, which deployed balloons into the atmosphere and landers onto the surface, as they flew past Venus on their way to encounters with Halley's comet in 1986.

Despite the mixed fortunes of the Soviet Mars programme, one should not judge it too harshly. The experience gained from it and from the Luna, Venera and VeGa series has allowed the Soviets to embark on their Phobos exploration programme (see page 18) and gave them the confidence to outline plans for the ultimate landing of men on Mars.

The first successful probe to Mars was launched by the United States in November 1964. For the next dozen years the US was to set the pace in the exploration of the planet.

In 1959, two years after Sputnik 1, NASA had given the Jet Propulsion Laboratory in Pasadena, California, the task of developing its unmanned planetary programme. JPL had built Explorer 1, the first US satellite; its next task was to build the Ranger series of spacecraft to explore the Moon and the Mariner series to investigate the inner planets.

The first two Mariners were sent to Venus: the second was successful and on 14 December 1962 became the first probe to fly past another planet. The next two Mariners were adapted for the trip to Mars and were ready by the time the next Mars launch window opened.

Mariner 3 was launched on 5 November 1964, with an Atlas-Agena combination as its booster. But the fibreglass shroud surrounding the spacecraft in the nose of the rocket failed to jettison.

Three weeks later, NASA tried again with Mariner 4, to which a

Past United States Missions

Year	Mission	Launch date	Arrival date	Comments
1964	Mariner 3	5 Nov	—	Launch failure.
	Mariner 4	28 Nov	14 Jul 65	First successful flyby and return of images.
1969	Mariner 6	25 Feb	31 Jul	Successful photographic flyby.
	Mariner 7	27 Mar	5 Aug	Successful photographic flyby.
1971	Mariner 8	8 May	—	Orbiter mission; booster failure.
	Mariner 9	30 May	14 Nov	Orbiter mission: mapped surface and photographed Martian satellites. Operated until 27 Oct 1972.
1975	Viking 1	20 Aug	19 Jun 76	Orbiter/lander mission. First soft landing (20 Jul) and pictures from the surface. Lander operated until 13 Nov 1982.
1975	Viking 2	9 Sep	7 Aug 76	Orbiter/lander mission. Soft landing (3 Sep) and pictures from the surface. Lander operated until 12 Apr 1980.

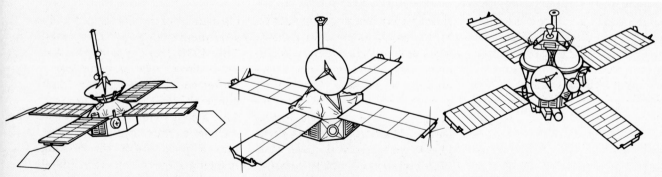

Mariner 4

Purpose: flyby
Flown: 1964–5
Launch mass: 575lb (261kg)
Span: 22.3ft (6.79m)
Solar panel area: 69.9ft^2 (6.5m^2)
Payload: TV system, radiation/magnetic field detectors

Mariner 6/7

Purpose: flyby
Flown: 1969
Launch mass: 910lb (413kg)
Span: 20ft (5.79m)
Solar panel area: 83.2ft^2 (7.74m^2)
Payload: TV cameras and scientific instruments for probing surface and atmosphere

Mariner 9

Purpose: orbiter
Flown: 1971–2
Launch mass: 2200lb (998kg)
Span: 22.6ft (6.9m)
Solar panel area: 82.8ft^2 (7.7m^2)
Payload: TV cameras and scientific instruments for probing surface and atmosphere

Launch of Mariner 7 on 27 March 1969 by Atlas-Centaur rocket.

'frames' to Earth. The technology of the time was such that Mariner 4 transmitted data at a rate of only 8.3 'bits' per second. As every frame required 240,000 bits, Mariner 4 took 10 days to send its pictures back to Earth. (A decade later, the Viking craft were to return their data 40,000 times faster.)

Surprising craters

Because of the geometry of Mariner 4's flyby, the spacecraft could view only a small strip of the Martian surface in the southern hemisphere—something like one per cent of its total surface area. In all, 11 Mariner 4 frames actually showed the surface. The seventh of these provided a shock: it clearly showed craters. A few planetary scientists had expected these, but the discovery came as a surprise to the general public.

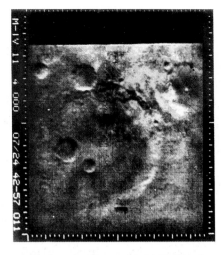

The 11th picture of Mars taken by Mariner 4, clearly showing craters.

Technicians work on the Mariner 4 spacecraft in preparation for launch.

more reliable metallic shroud had been added at the last minute. The launch on 28 November was successful and, despite problems with the craft's orientation sensors, Mariner 4 was injected into the correct transfer orbit for Mars.

After an eight-month journey, Mariner 4 flew past the red planet on 14 July 1965. It carried six scientific instruments, including a simple TV camera to observe the surface and return a total of 22

In early 1969 two more Mariners were launched towards Mars and arrived within a few days of each other, just a week after the end of the Apollo 11 mission. The craft were identical, each carrying more instruments than their predecessor. They encountered Mars over its southern hemisphere, observing about 20 per cent of the planet's surface area. The temperature of the south polar cap indicated that it was composed of carbon dioxide ice.

Viking 1/2

Purpose: orbiter/lander
Flown: 1975–82
Launch mass: 5132lb (2328kg) orbiter/1462lb (663 kg) lander
Span: 32ft (9.75m)
Solar panel area: 161ft² (15m²)
Payload: orbiter with TV cameras and surface thermal mapper; lander with TV, packages looking for life, analysing soil and atmosphere, and listening for Marsquakes.

American deep-space craft

NASA has launched four pairs of Mars probes, all based on an octagonal body, their sizes and capabilities increasing as more powerful rockets became available. The 1964 Mariner 4 was launched by the small Atlas-Agena D and could only fly past Mars carrying a small camera package. Nevertheless, it discovered that Mars was more Moonlike than previously believed, with craters and a very thin atmosphere. The larger Mariners 6 and 7 of 1969 included a more comprehensive science package but their Atlas-

Centaur rockets could still only send them sailing past the planet. The 1971 launch window was more favourable and the same rocket could despatch more than double the previous payload mass. This meant that Mariner 9 could include a retro-motor to brake into a polar orbit for a long mapping mission. In 1975 Viking could be about three times larger than Mariner 9 because it flew on the powerful Titan-Centaur booster, allowing not only a mapping orbiter but also a sophisticated lander to make detailed surveys including the search for life.

The new pictures again showed a Moonlike surface, with vast expanses of craters and unusually chaotic surface regions, named 'weird terrain' by the geologists.

It is one of the great ironies of planetary exploration that, by chance, these first Mariners missed the most exciting surface features. But better luck was in store for the next in the series, planned as the first Mars orbiters. The 1971 window was one of the best launch opportunities in this century, and at the start of the space age optimists had hoped that the first manned landing would take place then. Mariner 8 was launched on 8 May, but the autopilot on its Atlas-Centaur launcher malfunctioned and the rocket crashed in the Atlantic Ocean. The fault was rectified on Mariner 9, which was launched 22 days later.

The Viking lander's descent profile. The lander separated from the orbiter (top left), and at 21,000ft (6400m), the parachute was deployed. The lander was cushioned by small rockets. In this pre-mission painting the sky looks blue; in reality it turned out to be pink.

Shrouded in dust

While Mariner 9 was *en route*, astronomers observed the development of one of the fiercest Martian dust storms ever seen. By the time the probe entered Mars orbit on 14 November, the planet was engulfed by the storm. So its cameras were trained on the two Martian moons, Phobos and Deimos. They were revealed as small, irregularly shaped rocky bodies — probably captured asteroids.

It was only after several months that the dust began to dissipate and Mariner 9 could at last begin to investigate the planet. Over the next year it returned 7329 images, some disclosing surface details as small as 300 feet (100 metres) across. They revolutionized our view of the red planet.

To begin with, four large black spots were seen poking above the murky atmosphere. As the dust settled, they were clearly seen to be volcanoes. The largest was named Olympus Mons: it is the largest volcano that has been seen anywhere in the solar system. With continuing observation, a

This globe of Mars was made from over 1500 Mariner 9 photographs taken during 1971/2. Olympus Mons is visible at the equator, and the spiral pattern of the north polar cap is prominent.

'new Mars' was revealed, marked by vast canyons, dried-up water channels and oddly shaped features carved by the winds.

The ultimate aim of this first phase of Mars exploration was to land spacecraft on the surface. This was accomplished with the two Viking missions, each involving an orbiter carrying a lander. Launched in late 1975, Viking 1 reached Mars in June 1976 and was soon joined by the second, identical, craft in August.

On the Plain of Gold

Originally it had been hoped that the first Viking lander could touch down on 4 July, to coincide with the United States' bicentennial celebrations. Landing sites had been chosen from Mariner 9 images. But the Viking orbiters, equipped with better TV cameras and favoured with clearer skies, found that these sites were too rough. A desperate search for a smooth landing place was begun, and within two weeks one was found, at the western edge of Chryse Planitia—the Plain of Gold—near a region of dried-up water channels.

In the early hours of 20 July the first Viking lander separated from its orbiter and headed Marsward. The touchdown was successful, and soon the lander returned

striking pictures of a rock-strewn desert with a pink, hazy sky. Two months later the second Viking landed in a more northerly location known as Utopia, revealing much the same kind of landscape.

The three instruments on each orbiter and the 11 on each lander started to give us the most detailed pictures yet of another world. At the same time miniaturized onboard laboratories performed chemical and biological analyses on soil samples, searching for signs of life.

NASA had planned that each mission would last for 90 days. In fact Orbiters 1 and 2 operated for four and two years, respectively. Lander 2 finished operations in 1980, while Lander 1 continued returning information for over six years, until contact was lost at the end of 1982.

Left: a full-size model of the Viking lander, its sampling arm extended. In the Viking 2 surface view (below), the radio dish and nuclear power source (on the left) are conspicuous.

THE PHOBOS MISSION

Over the next decade, the Soviet Union is planning the most ambitious unmanned missions to Mars ever attempted. Mars is the focus for Soviet planetary exploration: several launch windows from now until 2000 will see spacecraft launched towards the planet from Tyuratam. The start of the programme is the Phobos mission to explore at least one Martian moon at close range; by the end of the century roving vehicles will have explored the Martian surface and returned samples of the planet's soil to Earth.

International cooperation

Remarkably, the Soviets have revealed much of their planning quite openly to their European and American colleagues. They are so confident of their lead that they have even offered 'international collaboration' on an unprecedented scale.

International cooperation is not new in the Soviet unmanned space programme, but it has mainly been with the French. As long ago as 1971, a French experiment was carried on the Mars 3 orbiter. It was a French proposal that inspired the balloons dropped into the atmosphere of Venus by the VeGa spacecraft in 1985. And the Russians went so far as to put a US dust detector on the VeGas for their subsequent flyby of Halley's comet.

The highly successful VeGa craft gave proof that Soviet space technology has overcome previous problems and improved to the point where it can handle long-duration unmanned missions.

Even so, Soviet engineers do not have the full panoply of technology available to the West. Though the Phobos mission probes are their most complex to date, the basic design of the craft can be traced back to the earlier

Mars series. Original Soviet studies envisaged the whole Phobos spacecraft actually landing on the satellite and returning a sample to Earth. But this was judged too ambitious, and so the present plan evolved: that the craft should hover above Phobos and despatch smaller landers to its surface.

In accordance with the Soviets' traditional preference for dual missions, the Phobos mission comprises two spacecraft. The idea is simple: if one craft fails, the other can carry out most of the mission objectives. If both craft function properly, each can corroborate the other's findings.

Encounter with Phobos

Two Proton boosters will launch the two Phobos craft towards Mars within a few days of each other. Each spacecraft is very heavy (around 6 tons) and will need to fire its onboard propulsion system to assist its escape from the Earth. Each spacecraft carries numerous experiments to investigate the planet, its two moons and the interplanetary environment. Those instruments are listed, along with the countries collaborating on their construction, on pages 22/3. Many have been provided by more than one country, under Soviet supervision. A number of the scientific experiments are very versatile: for example, the TV cameras used to observe Phobos in detail will be trained on Mars before and after the close encounter with the satellite. On their arrival after six months' flight, the craft will go into elliptical parking orbits around Mars.

In order to achieve a close encounter with Phobos, the spacecraft must carry out a long series of measurements of its position as it moves around Mars, refining the inadequate data on its orbit currently available. Over a period of nearly three months, each Phobos

craft will alter its orbit no less than four times to achieve close encounter with the moon (see page 20). Then the first craft will sweep to within 150 feet (50 metres) of Phobos, no mean feat of interplanetary navigation. For about 20 minutes it will 'hover' over the moon, tracking systematically across the surface.

It will analyse the composition of the satellite's surface (see boxes on pages 22–4), and drop two landers. Because the moon is so tiny and its gravity so weak, there is a real danger that they will bounce off the surface. The larger of the pair, called the Long-Term Automated Lander, carries a harpoonlike device with which to anchor itself to the surface. It will function for about a year, taking television pictures and further analysing the surface material.

The smaller lander will hop across Phobos by means of powerful springs. Not surprisingly, the Russians call it 'the frog'. If the first Phobos encounter is successful, it is just possible that the second spacecraft may be directed to Deimos; but Soviet scientists have admitted that the extensive modifications of the spacecraft's computer software that would be required to enable it to hover over Deimos may preclude this. The main spacecraft will continue to orbit Mars for another two years. During this time the Phobos craft will be able to observe the Sun from the side opposite to the Earth. This unique vantage point will give solar scientists the chance to monitor the effect of the Sun's activity in interplanetary space with unprecedented accuracy.

This concentration on Phobos is believed to signal a growing Soviet interest in making this tiny moon the site of a manned landing. It could be an ideal spot to set up a base camp from which to explore Mars itself.

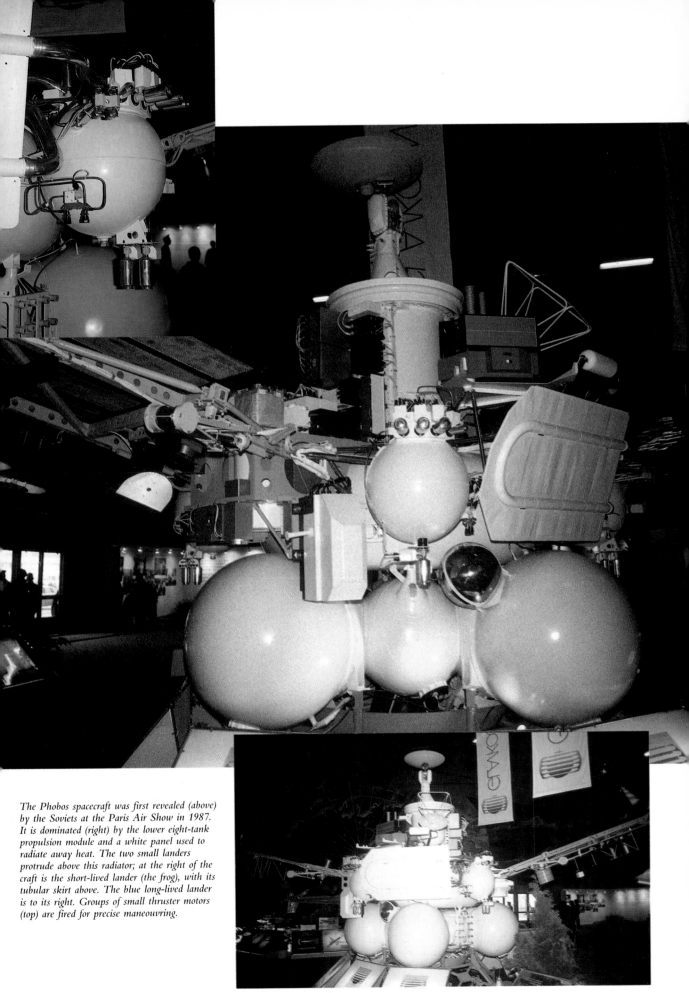

The Phobos spacecraft was first revealed (above) by the Soviets at the Paris Air Show in 1987. It is dominated (right) by the lower eight-tank propulsion module and a white panel used to radiate away heat. The two small landers protrude above this radiator; at the right of the craft is the short-lived lander (the frog), with its tubular skirt above. The blue long-lived lander is to its right. Groups of small thruster motors (top) are fired for precise maneouvring.

The Journey to Mars

The Phobos mission spacecraft will be launched from Tyuratam into a parking orbit almost 120 miles (200 kilometres) above the Earth, tilted at 51.5° to the equator. About an hour later, the Proton fourth stage will ignite to increase the speed by 2–2.5 miles per second (3.5–4 kilometres per second), flinging Phobos outwards to begin its trek towards Mars.

At the beginning of the 200-day transit, engineers will activate each system on Phobos in turn to check the craft's health, and will take any necessary remedial actions. But vigilance will be necessary throughout the mission: all long-lived planetary spacecraft develop personal quirks requiring patient nursing.

Since the batteries would discharge within days if their power were not replaced, Phobos must lock its long axis onto the Sun so that its solar panels can generate electricity.

After 7–10 days, ground-based radio measurements, assisted by onboard equipment, will have measured the trajectory precisely, and engineers will know whether the main engine has to be ignited for a brief course correction.

The cruise to Mars allows uninterrupted scrutiny of the Sun by an array of telescopes and other instruments, while the fields and particles experiments measure the radiations and electromagnetic fields in interplanetary space.

The Earth, being closer to the Sun, races ahead of the Phobos spacecraft. Radio commands will gradually take longer to reach the craft, the time stretching to 10.5 minutes by the time of Mars arrival. If Earth contact is lost, Phobos will search for the singal, under the control of the onboard computer's preprogrammed instructions. Some 7–15 days before arrival, a second course correction could be made with the main engine to 'tweak' the exact point of arrival.

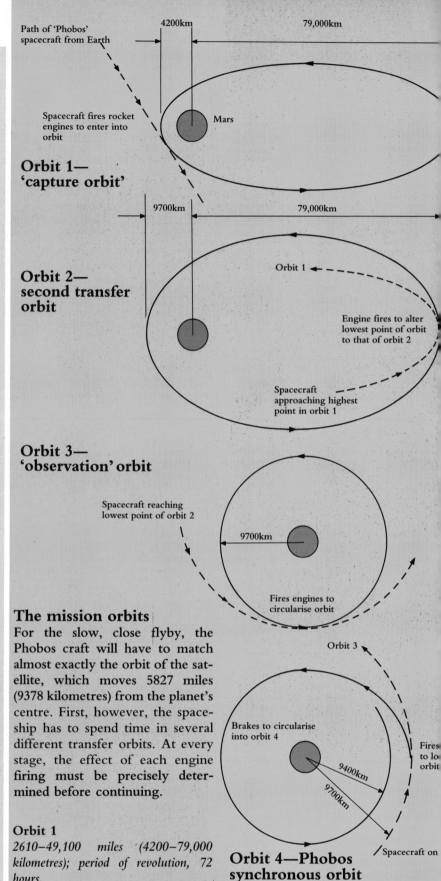

Orbit 1— 'capture orbit'

Orbit 2— second transfer orbit

Orbit 3— 'observation' orbit

Orbit 4—Phobos synchronous orbit

The mission orbits

For the slow, close flyby, the Phobos craft will have to match almost exactly the orbit of the satellite, which moves 5827 miles (9378 kilometres) from the planet's centre. First, however, the spaceship has to spend time in several different transfer orbits. At every stage, the effect of each engine firing must be precisely determined before continuing.

Orbit 1

2610–49,100 miles (4200–79,000 kilometres); period of revolution, 72 hours.

When the two Phobos craft arrive at Mars, they disappear behind the planet. Near the point of closest approach, 500 miles (800 kilometres) above Mars, they ignite their main engines to kill off some of their speed, allowing the planet's gravity to capture them in high, looping orbits. They remain in these orbits for 25 days, making high-resolution studies of the surface as they sweep through the lower parts of the orbits, 500 miles (800 kilometres) above the surface.

Orbit 2

6030–49,100 miles (9700–79,000 kilometres); period of revolution 79 hours.

The spacecraft heading for Phobos raises its orbital low point by means of an engine burn at the orbit's *high* point, rather like pressing on one end of a lever to produce an effect at the other end. At this stage the new orbit is made equatorial. The craft stays for 35 days in this orbit, which is completely outside the satellite's. Then it brakes at its low point, lowering the far point, to create a new orbit.

Orbit 3

6030 miles (9700 kilometres) circular; period of revolution 8 hours.

The new orbit is circular and passes just 200 miles (320 kilometres) above its intended final orbit, which matches that of the satellite. The moon's path is currently too poorly known from ground-based observations for a direct approach; photographing it against the background of the stars during this important 30-day period will provide ground controllers with the necessary navigational information.

Orbit 4

5827 miles (9378 kilometres) circular; period of revolution 7.6 hours.

After the main propulsion module has been jettisoned, the hydrazine thrusters slot the spacecraft into a satellite-matching path by means of braking burns on opposite sides of the orbit. The final approach can now begin.

After its close approach, the Phobos craft fires its thrusters to return to its previous eight-hour circular orbit (orbit 3) for at least 140 days of Mars observations. If this first mission is successful the second craft is likely to follow to make corroborating observations.

Final approach and hovering

For the final approach, the Phobos craft remains in an orbit up to 35 miles (60km) from the satellite, its solar cells still pointing towards the Sun. Then the attitude-control system turns the craft and locks onto the moon (1, 2). Braking thrusters (3) bring it down to within 1.25 miles (2km) of the surface (4). The aim is to approach to within

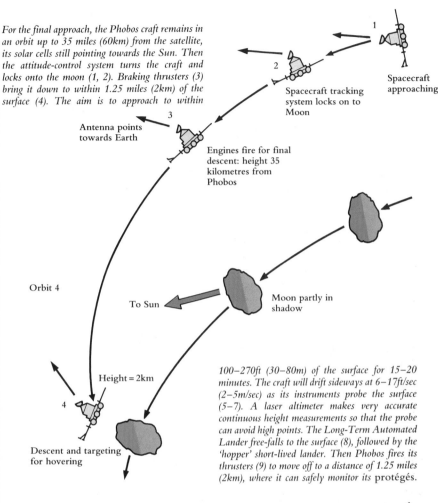

100–270ft (30–80m) of the surface for 15–20 minutes. The craft will drift sideways at 6–17ft/sec (2–5m/sec) as its instruments probe the surface (5–7). A laser altimeter makes very accurate continuous height measurements so that the probe can avoid high points. The Long-Term Automated Lander free-falls to the surface (8), followed by the 'hopper' short-lived lander. Then Phobos fires its thrusters (9) to move off to a distance of 1.25 miles (2km), where it can safely monitor its protégés.

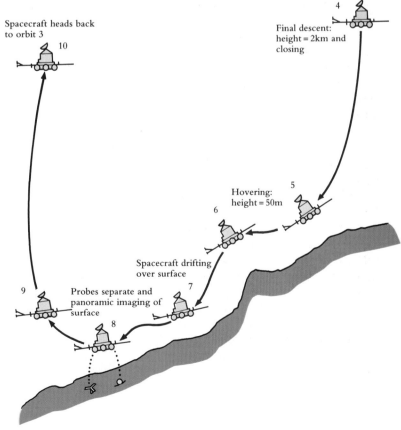

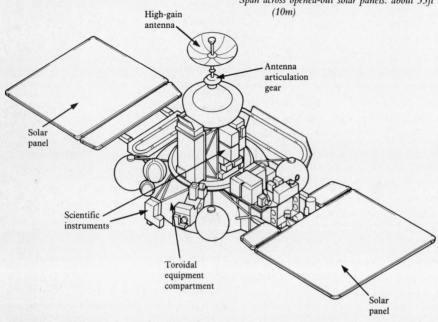

The Phobos spacecraft
Mass: 6 tonnes
Height: about 11.5ft (3.5m) overall
about 8ft (2.5m) excluding propulsion
module
Width: about 11.ft (3.5m) (propulsion module)
Span across opened-out solar panels: about 33ft
(10m)

High-gain
antenna

Antenna
articulation
gear

Solar
panel

Scientific
instruments

Toroidal
equipment
compartment

Solar
panel

The design of the Phobos craft

Phobos is the first of a new generation of Soviet deep-space craft, but its design derives extensively from the old Mars and Venera vehicles. The central cylindrical structure and the torus at the base have been retained as protective pressurized and temperature-controlled enclosures for the delicate computer control and communications electronics. The scientific instruments are mounted externally, where they enjoy clear fields of view. The solar panels convert sunlight into electricity to be routed into internal storage batteries. Excess internal heat is radiated away by two large white panels. Radar equipment for controlling the close approach, in conjunction with a laser altimeter, is mounted under the solar panels.

Older Soviet planetary craft suffered from reliability problems, but Phobos incorporates new command and radio systems, including a computer with a 30-megabit memory. At the satellite

the craft will be too distant from Earth for real-time control, so the computer is preprogrammed to execute set routines, modified by information provided by the altimeter and other instruments. Data will be sent to Earth at 4000 bits per second—the rate at which the American Viking Mars landers transmitted.

The lower propulsion module carries an improved Venera engine with a thrust of 1–1.9 tons, used for major manoeuvres. It draws on eight aluminium-alloy spherical tanks of nitric acid and an amine-based fuel. This engine is used for the large braking burn that puts the craft into Mars orbit, and the module is then jettisoned. Smaller hydrazine thrusters are mounted on the upper stage and fed from four spheres mounted on the torus. Four of these engines deliver one pound (0.5 kilogram) of thrust, while the remaining 24 deliver 11 pounds (five kilograms). The network of thrusters will be brought into play for the final manoeuvres towards the satellite.

The Phobos orbiter's scientific instruments

FREGAT
Surface mapping by TV cameras and spectrometer (Phobos, Mars)
USSR, East Germany, Bulgaria
Three linked TV cameras and a spectrometer (which analyses light into its constituent wavelengths) will provide the most spectacular close-ups and panoramic views of Mars and Phobos. From 150 feet (50 metres) above Phobos they will reveal details as small as 2.5 inches (six centimetres).

As these images are recorded, the spectrometer will break up the light into 36 wavelength bands; the subtle changes in colour thus revealed will show up variations in mineral composition across the surface.

The craft's computer can store 1100 groups of pictures from the four instruments. There is a 60 per cent overlap between pictures, allowing stereoscopic images to be formed. The pictures will track precisely the same areas as those covered by the other instruments described here. The cameras will also help navigation by photographing Phobos against star backgrounds to pinpoint it.

LIMA-D
Soil analysis by laser (Phobos)
USSR, Austria, Bulgaria, Czechoslovakia, East and West Germany.
Since the main spacecraft will not land, scientists need to discover the soil's composition from observations made at least 100 feet (30 metres) above it. Laser pulses will vaporize a spot of soil 1/20–1/10 inch (1–2 millimetres) across and up to a ten-thousandth of an inch (2 micrometres) thick. From the cloud of vapour that rises from these, LIMA-D will capture ions (atoms that have lost or gained electrons and hence become electrically charged) at the rate of a million every 5–10 seconds. A mass spectrometer will sort these by mass to reveal the quantities

of elements present, ranging from hydrogen to lead.

DION
Soil analysis by ion beam (Phobos)
France
Complementing LIMA-D, DION will fire energetic krypton ions to knock further ions out of the top twentyfive-thousandth of an inch (one micrometre) of the soil, again for sorting by a mass spectrometer into the spread of elements from hydrogen to nickel. The detector is also sensitive enough to pick up particles liberated by the solar wind, the stream of ions from the Sun that bombards Phobos continually.

GRUNT
Radar surface mapping, probing of internal structure (Phobos)
USSR
Radar transmitters and receivers mounted below the solar panels will operate at three radio frequencies to penetrate more than 650 feet (200 metres) deep. The reflections of signals at 500 MHz and 130 MHz will map the surface at resolutions of 1.3 feet (0.4 metres) and 6.5 feet (2 metres) respectively, and indicate the surface materials' electrical properties. They will also indicate the rock density 100–650 feet (30–200 metres) down. A 5-MHz frequency has been added to delve even deeper, but its resolution is only 500 feet (150 metres).

An experiment called PLAZMA is also carried, which uses radar signals of frequency 0.18–3 MHz to probe the Martian ionosphere (a layer of the outer atmosphere).

KRFM-ISM and TERMOSKAN
Broad temperature and mineral surface mapping in the infrared (Phobos, Mars)
USSR and France / USSR
These instruments will extend FREGAT's mineral studies to longer wavelengths and add surface temperature measurements. Observations in certain infrared wavelength bands will generate temperature maps, revealing seasonal variations in 'hot spots' and permafrost on Mars.

GS-14
Chemical element mapping by X-rays (Phobos, Mars)
USSR
Some naturally radioactive elements betray their presence by emitting characteristic gamma rays, while other elements are stimulated to emit them when bombarded by cosmic rays. GS-14 detects these telltale emissions by means of a caesium iodide crystal. It can detect radioactive uranium, thorium and potassium, and basic rock-forming elements such as oxygen, silicon and iron. GS-14 will construct maps of the distribution of these elements, with a resolution for Phobos of 2.5 miles (four kilometres).

IPNM-3
Chemical element mapping by neutrons (Phobos)
USSR
Similar in principle to GS-14, this instrument detects characteristic neutrons emitted by oxygen, aluminium, iron etc, when excited by cosmic rays. It will measure the surface layer's water content.

AUGUST
Atmospheric composition with height (Mars)
France, USSR
AUGUST will measure the sunlight passing through the Martian atmosphere at dawn and dusk to discover how much is absorbed at various wavelengths characteristic of the air's constituents. Temperature, pressure and dust content will also be revealed.

ASPERA
Maps particles in magnetosphere of Mars (Phobos, Mars, Sun)
Finland, Sweden, USSR
ASPERA is a mass spectrometer measuring the energies and masses of ions and electrons in and around the magnetosphere of Mars—the region where these charged particles, streaming from the Sun in the solar wind, interact with the planet's magnetic field. At Phobos ASPERA will function as part of the DION experiment. In addition, experiments called SOVKOMS, TAUS and HARP will extend the range of charged-particle studies.

MAGMA and FGMM
Magnetic field measurements (Phobos, Mars, Sun)
USSR
Measurements will be made of magnetic fields around Mars and Phobos and in interplanetary space on the voyage, to gain insight into their structures.

APV-F
Plasma wave measurements (Phobos, Mars, Sun)
USSR, Czechoslovakia, ESA, Poland
APV-F will map the magnetic and electric fields and their variations associated with the charged particles ('plasma') measured by ASPERA.

SOLAR STUDIES
Observations of the Sun during voyage and after Phobos flyby
Three telescopes will observe 'soft' (low-energy) and ultrasoft X-rays and visible light. Sometimes they will be switched on automatically when the RP-15 device detects increased solar activity. These telescopes will examine the Sun's surface and outer layers, and will even provide stereoscopic views. Two experiments called LET and SLET will monitor solar cosmic rays in conjunction. IPHIR will monitor light output at three wavelengths, one in the ultraviolet, one in the green and one in the infrared. It will be looking for solar oscillations—an expanding movement of the Sun that would give clues to its internal structure and activity. Finally the LILAS and VGS instruments will monitor solar gamma-ray output and look for the mysterious gamma-ray bursts that occasionally sweep across the solar system.

The Long-Term Automated Lander

Towards the end of its 15–20-minute hover above Phobos, each spacecraft jettisons two small landers to make measurements at ground level. The first to depart is the Long-Term Automated Lander, which makes contact with the surface at a speed of about three feet per second (one metre per second). As it does so, it fires a harpoonlike penetrator into the surface as an anchor, since Phobos's gravity is too weak to hold the lander down firmly.

An electric motor winches a tether tight and holds the lander firmly against the surface while the main instrument platform rises by 30 inches (80 centimetres) to give the TV cameras a better view.

Once any dust has settled, three petal-like solar panels unfold to generate electricity for the internal batteries, which will keep the instruments powered for up to a year. An optical sensor keeps the panels pointing sunwards. The onboard scientific instruments photograph and analyse the soil, and listen for seismic tremors.

The craft also acts as a radio beacon for a network of large radio telescopes in the USSR and USA, enabling them to measure the Earth–Phobos distance of scores of millions of miles to an accuracy of 15 feet (five metres). Such measurements could confirm suspicions that Phobos is slowly spiralling in to destruction on Mars in 100 million years.

LTAL instruments
ALPHA-X
Soil composition (Phobos)
USSR, West Germany

In one part of the experiment, a radioactive curium-244 source irradiates the surface with alpha particles and the radiation that is emitted by the surface is measured to reveal the concentrations of elements present, ranging from beryllium to iron. The second part of the experiment looks for the characteristic X-rays emitted by elements heavier than sodium in response to irradiation by curium-244 and cadmium-109. ALPHA-X will thus evaluate the composition of the surface material.

RAZREZ
Surface layer properties
USSR

Accelerometers in the penetrator record its deceleration on impact, indicating the surface rock type. Thermometers provide information from below the surface.

SEISMOMETER
Surface disturbances
USSR

As on Earth, seismometers measure surface movements to help geologists understand the body's internal structure. If two long-lived landers are placed on Phobos, they will be able to pinpoint the quakes' origins more precisely.

TV SYSTEM
Surface imaging
USSR, France

Two wide-angle cameras, each having a field of view of $36° \times 27°$, will provide stereoscopic views of the entire landing area as the platform rotates to follow the Sun.

SUN SENSOR
Measurements of Sun's position
USSR, France

Observing the Sun's precise position from both the long-lived landers will determine the satellite's libration (a slight nodding motion as it rotates)—an indicator of its internal mass distribution.

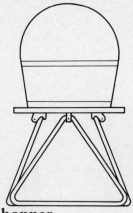

The hopper

The second lander to be launched is the short-lived 'hopper', designed to wander over the surface measuring soil strength and composition, and local magnetic and gravity fields. It is 20 inches (50 centimetres) across and weighs 45–65 pounds (20–30 kilograms).

A protective skirt is released and four battery-powered rods turn the hopper over so that its flat instrumented side is against the ground. When it has completed its series of measurements at that spot, it hops as much as 65 feet (20 metres) sideways to begin a new cycle. The hop is quite easy since Phobos's gravity strength is a thousandth of the Earth's.

Hopper instruments
All USSR
X-RAY FLUORESCENCE SPECTROMETER
Analyses the chemical composition of the soil.

PENETROMETER
Records soil strength by means of a device pushed into the surface.

DYNAMOMETER
Measures deceleration on impact after each hop, giving indication of surface structure.

MAGNETOMETER
Measures the satellite's magnetic field, which gives indications of the internal rock structure.

GRAVIMETER
Measures the local gravity field.

The main Phobos spacecraft hovers above the satellite, having fired its powerful LIMA-D laser at the surface. The Long-Term Automated Lander has touched down on the surface; the hopper is still slowly falling towards the moon.

AFTER PHOBOS

Beyond the Phobos mission, Soviet plans are by no means firm, with scientific priorities still being determined. However, their plans for unmanned missions to Mars for the 1990s are, in outline:

1) Select landing sites.
2) Send rovers across the surface.
3) Return soil samples to the Earth.

A mission is planned for the 1992 launch window, but the Soviets have said that it might be delayed until 1994 to give more time to digest the results of the Phobos mission. Two orbiters will be placed in highly elliptical orbits passing almost over the poles of Mars. As the planet rotates, they will in time be able to build up a picture of the whole surface. High-definition TV cameras, having a resolution of 30 feet (10 metres), together with highly sensitive radar, will be employed to map possible landing sites.

Each orbiter will carry a balloon, a small rover, and possibly penetrators. The penetrators will be small, shell-like devices, fired from orbit or dropped from a balloon. They will pierce the surface and analyse the chemical composition of the bedrock. They may also carry radio beacons to provide navigational fixes for future rovers.

The balloons, which might be provided by the French, would be an ingenious way of exploring the planet. With an Earth weight of 110 pounds (50 kilograms), each would be able to survive for up to six weeks in the thin carbon dioxide atmosphere. By day the Sun would warm the atmospheric gas inside the open-bottomed balloon, causing it to rise to 2.5–3.5 miles (4–5.5 kilometres). It would drift as much as 120 miles (200 kilometres) each day. Its altitude could be controlled by opening a valve at the top to release warm gas. The surface could be photo-graphed from the balloon with staggering clarity. At night the gas would cool and the balloon would descend to make chemical analyses of the soil. It would be suspended just clear of the hazardous surface by an upper helium balloon of 140,000 cubic feet (4000 cubic metres) volume. Each balloon would be tracked by means of its radio transmissions, giving information about currents within the Martian atmosphere.

The Marsokhods

In one projected design, each rover, or Marsokhod, would have six wheels, would weigh 330 pounds (150 kilograms), and would be nuclear-powered. It could wander for as much as 180 miles (300 kilometres) over the surface, making detailed soil analyses and weather measurements, and searching for biological activity with its 44-pound (20-kilogram) science package.

If this mission goes ahead, as hoped, in 1992, a separate mission in 1994 could land two large rovers on Mars to traverse the surface for up to a year. Each would be capable of digging 65–100 feet (20–30 metres) into the surface. One would investigate the chemical and physical properties of the cores thus obtained; the other would test for signs of life. The Soviets freely admit their belief that there *is* life on Mars.

It is possible that in 1996 the Soviets, drawing on their by-then extensive Mars experience, will attempt a sample return mission. It will include both a return vehicle and a rover to wander over the surface looking for suitable samples. Whereas NASA is considering flying these elements separately (see page 37), the Soviets prefer the idea of a combined descent vehicle. That vehicle would first aerobrake into Mars orbit and then descend to a site determined from the previous missions. What would happen then is still under study, but scientists agree that the rover would move away from the landing area, contaminated by the rocket exhaust, to scoop and drill for several pounds of samples. The rover could then either return to the lander ascent section or, in a beefier version, despatch its own return vehicle. The rover could continue its work by remote control long after the ascent stage had departed with its precious samples.

When the time comes for return, the ascent vehicle might head directly for Earth, but this would require a large lander, and the whole potentially contaminated vehicle would be delivered to the Earth. Instead, the ascent stage will probably dock with a Mars orbiter and transfer its samples into a small, specialized Earth return vehicle. In turn that could either enter Earth's atmosphere directly or brake to enter a safe orbit, where a space station could conduct biological and quarantine processing.

If it proves feasible to fly this mission in 1996, then a second pair could depart two years later. A manned flyby of Mars could pick up samples from the ascent stages of these craft at the turn of the century – a possibility that cannot be dismissed. The Soviets are also talking of a long-lived rover with a 600-mile (1000-kilometre) range, departing in the 2002 launch window. By this stage there could be men present to drive the rover—though they would be thousands of miles away, circling in Mars orbit or stationed on Phobos. From there they would be able to direct the rover by 'telepresence' (remote control), unhampered by the delay of several minutes that affects Earth-Mars communication.

The strange case of the Vesta mission

An idea of how much the Soviets' planetary programme has changed to emphasize Mars may be gleaned from the twists and turns of their Vesta mission plans. These were first revealed to the West at an annual gathering of planetary scientists in Houston in March 1985. Delegates were astounded by the openness of their Soviet colleagues. For the first time they announced missions—Lunar Orbiter and Vesta—that had not yet been officially approved within the Soviet Union.

The Vesta mission was the subject of a joint study between the Soviet Academy of Sciences and the French Centre National d'Études Spatiales (CNES). It involved the concept of two spacecraft in one: they would be launched together but would separate, with the Soviet half going to Venus and the French part going to the asteroid Vesta. In keeping with Soviet practice, two of the combined spacecraft would be launched during the same launch window.

Vesta is the third largest asteroid, and a prime target for scientific investigation. Together with comets, asteroids are known as 'small bodies' and are among the more enigmatic members of the solar system. They have changed little since their formation, and may hold important clues about the origin of the solar system. The French Vesta craft would fly past several asteroids and possibly one comet, firing a penetrator at one of the asteroids.

But by the time the mission was officially approved, in October 1985, the thrust of the Soviet planetary programme had changed: the Soviet spacecraft would head towards Mars. The joint Franco-Russian design study was expanded to allow European involvement through the European Space Agency (ESA). The Soviet scientists were keen to return to Mars as soon as

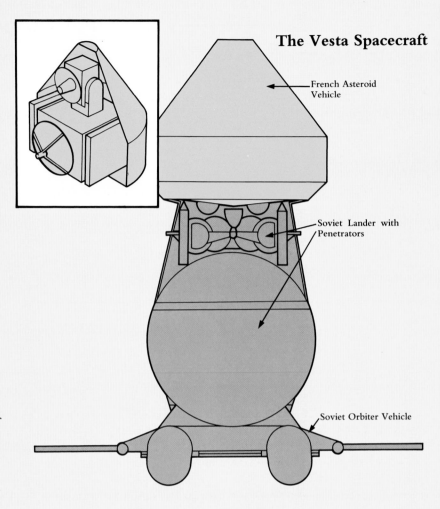

The Vesta Spacecraft

French Asteroid Vehicle

Soviet Lander with Penetrators

Soviet Orbiter Vehicle

possible—which would be during the 1992 launch window. Their CNES colleagues realized that their spacecraft would not be ready by then. So a contingency plan was drawn up: the Soviet Mars orbiters would be launched in 1992, while the French Vesta craft would leave in 1994.

At the 1987 Houston meeting the Soviet delegates announced the advanced plans for Mars exploration outlined on these pages. But the exact details have yet to be

worked out, and the first Mars orbiters may be launched in either 1992 or 1994.

To complicate matters further, ESA and CNES have been studying additional Vesta mission options. If the Soviet Mars orbiters don't fly until 1994, they may or may not be launched with the French small bodies spacecraft on the same Proton vehicles, for ballistic reasons. The final decision won't be taken by the Soviets, CNES and ESA until the end of 1988.

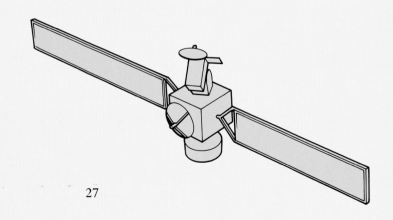

Year	Mission	Launch date	Time to Mars	Comments
1992	Lander	Sep/Oct	350 days	Craft enters polar/Sun synchronous orbit before lander descends to surface. Exploration by balloon and rover.
1994	Lander	Oct	315 days	Craft enters polar orbit before lander descends to surface. Exploration for up to a year by rovers.
1996	Sample return	Nov	315 days	Craft enters polar orbit before lander descends to surface. Soil-sampling rover returns specimens to ascent stage for return to Earth.

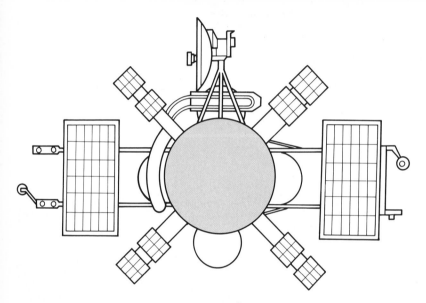

1992/94 MARS PROBE. A small rover and balloon package are housed inside an aerodynamic descent cone, 7ft (2.2m) wide and 11ft (3.3m) high. The cone is mounted on top of a Phobos-type propulsion module and equipment bay. The overall size of the probe is similar to the Phobos craft. Once the cone has been sealed on Earth, it will hold terrestrial bacteria at bay. During entry into the Martian atmosphere it will provide thermal protection for the equipment inside. The solar panels are stored upright for launch but once in space they swing down into their operating positions.

Mars orbiter

Descending module with Mars rover

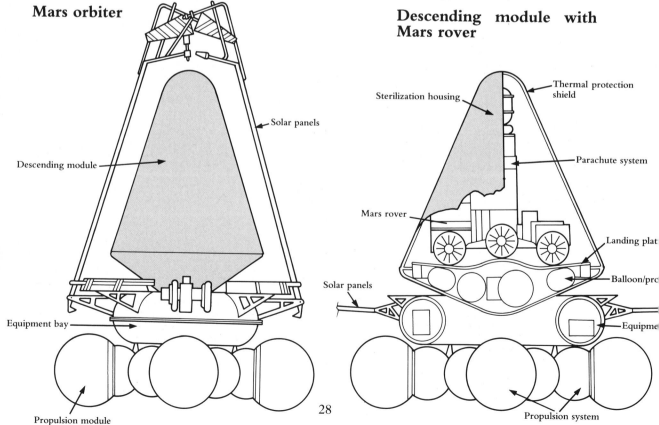

Solar panels

Descending module

Equipment bay

Propulsion module

Sterilization housing

Thermal protection shield

Parachute system

Mars rover

Landing plat

Solar panels

Balloon/pro

Equipme

Propulsion system

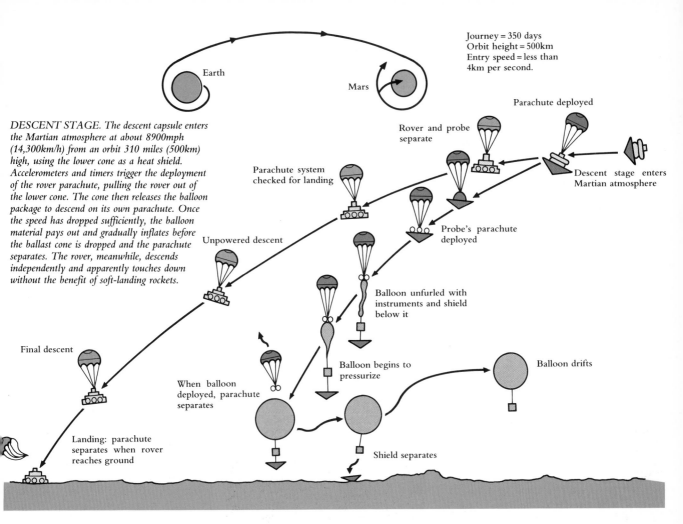

Journey = 350 days
Orbit height = 500km
Entry speed = less than 4km per second.

Earth

Mars

Parachute deployed

Rover and probe separate

Descent stage enters Martian atmosphere

Parachute system checked for landing

DESCENT STAGE. The descent capsule enters the Martian atmosphere at about 8900mph (14,300km/h) from an orbit 310 miles (500km) high, using the lower cone as a heat shield. Accelerometers and timers trigger the deployment of the rover parachute, pulling the rover out of the lower cone. The cone then releases the balloon package to descend on its own parachute. Once the speed has dropped sufficiently, the balloon material pays out and gradually inflates before the ballast cone is dropped and the parachute separates. The rover, meanwhile, descends independently and apparently touches down without the benefit of soft-landing rockets.

Probe's parachute deployed

Unpowered descent

Balloon unfurled with instruments and shield below it

Final descent

Balloon begins to pressurize

Balloon drifts

When balloon deployed, parachute separates

Landing: parachute separates when rover reaches ground

Shield separates

Mars rover

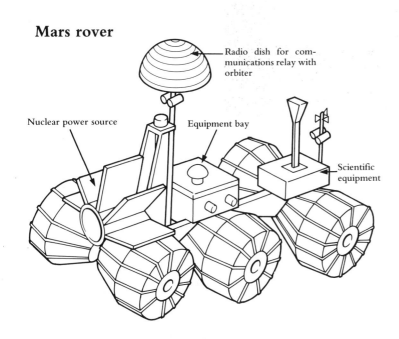

Radio dish for communications relay with orbiter

Nuclear power source

Equipment bay

Scientific equipment

MARS ROVER. This design, under consideration for the 1992/94 mission, has six wheels and weighs 330lb (150kg). It is powered by electricity generated thermoelectrically by heat from a radioactive source. The wheels have independent suspensions and drive motors to ensure maximum mobility over the terrain. Communications, control and scientific equipment is housed in sealed thermally regulated compartments for protection against the extremes of Martian temperature. The instruments will probably include a soil drill, soil and atmospheric gas analysers, a seismometer, and gauges for wind speed and atmospheric pressure and temperature. These will be contained in a 'science package' with an Earth weight of 44lb (20kg).

In this artist's impression, a Soviet sample return mission lander is descending towards the Martian surface. An atmospheric balloon has unfurled and pressurized above it, the parachute having already been jettisoned.

The United States committed itself to a Mars expedition on the day that Apollo 11 was launched towards the first manned landing on the Moon—16 July 1969.

President Nixon, who favoured a vigorous space programme, had set up a Space Task Group as soon as he took office. He had told the Group to advise him on what the nation's space plans should be after Apollo. NASA already had plans to investigate the Martian surface with unmanned Viking craft—now the Task Group proposed to send men there. It was Vice-President Spiro Agnew who announced the new national goal: a manned landing on Mars by the year 2000.

Its most enthusiastic and articulate advocate was the then NASA Administrator, Dr Thomas Paine, whose outspokenness earned him the reputation of being a 'swashbuckler'. (He is just as vigorous today as one of the leaders of the 'Mars Underground'.) Two months after Agnew's announcement came an even more exciting forecast, in the Task Group's official report. It called for an intensive programme that could put men on Mars by 1983, or 1986 at the latest.

Yet when those deadlines came and went, not only was there no American manned Mars programme—even the commitment to one had been forgotten. What had gone wrong?

The trouble stemmed in part from an unexpected, almost unbelievable, public boredom with space endeavours—a blasé attitude that had set in long before the Apollo programme came to an end. The nation that had gone wild with pride at the achievement of landing men on the Moon was asking: 'Why do we keep going back there? We've done what we said we would do.'

And the war in Vietnam and the economic problems it was causing were foremost in the public mind. When the cost of the Task Group's ambitious plan was put to President Nixon he balked at it. When Spiro Agnew, at a formal dinner, again spoke of sending men to Mars he was publicly booed.

The rot had set in. The Soviet Union's resolve to re-establish its lead in space must have been given a powerful boost at this point.

The Shuttle

The 1969 plan to send men to Mars had envisaged two main stepping stones on the way: a large space station and a reusable space ferry, or Shuttle, to build it. The only part of the plan that survived was the development of the Shuttle. Even so, when NASA's Marshall Space Flight Center in Alabama put a price tag of $10 billion on it, Congress refused to approve the cost. For the first time NASA was given a limit to its start-up budget on a major project: $5 billion.

So a compromise evolved: the Shuttle would be only partially reusable. Its expendable components would be a large external fuel tank and a couple of strap-on solid-fuel boosters. Such boosters were not favoured, particularly for manned flight, since they were unreliable. But they were one way of keeping down costs.

It took longer than expected to develop the Space Shuttle, and when it first took off from Cape Canaveral in April 1981 it was three years late. By this time there were many who said it was an idea behind its time. After a couple of years, the fulfilment of the original promise of one flight per week looked a long way off.

The Space Station

In his State of the Union message in 1984 President Reagan announced that he had directed NASA to develop a manned space station and to have it in Earth orbit by 1992. But the public showed little interest, and the Department of Defense said it had no use for it. So NASA sought the cooperation of foreign partners, such as Europe, Canada and Japan, and they agreed to participate.

But still there were no long-term plans for space activity. The following year Congress set up a National Commission on Space to advise on long-term goals, with special emphasis on the next 30 years. President Reagan appointed the members and he chose Tom Paine to chair it. Neil Armstrong was a member, and so was the Shuttle astronaut Kathryn Sullivan. It looked as if someone meant business: the Commission was told to report within the year.

By the time it did so, tragedy had struck the American space programme: the Shuttle *Challenger* was lost on 28 January 1986 with the death of seven astronauts. This event overshadowed the Commission's report, melodramatically entitled *Pioneering the Space Frontier*, which was published in August that year. It urged a manned

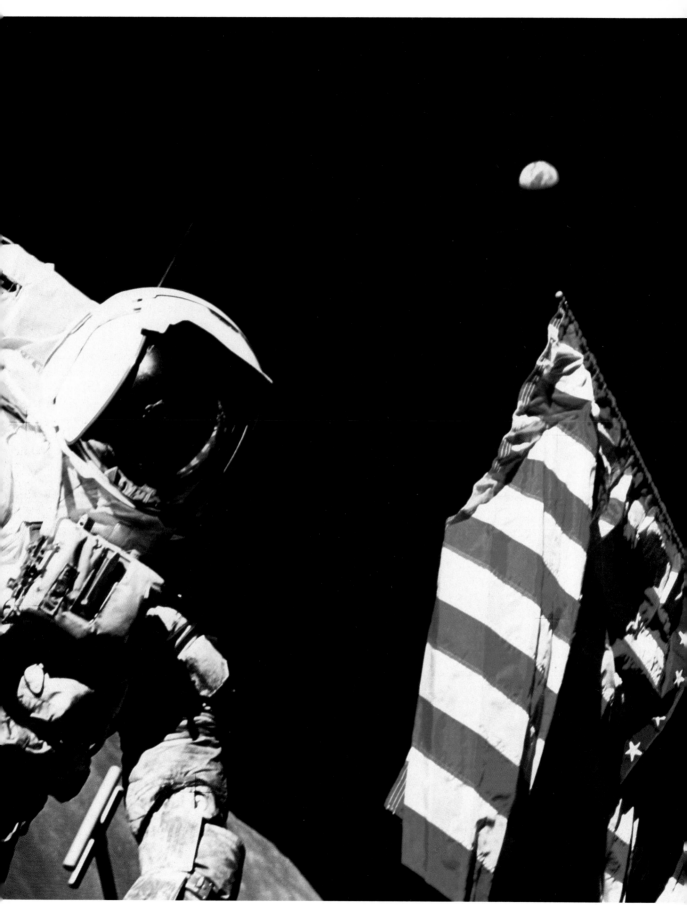

The distant Earth appears above the US flag, planted on the lunar surface in December 1972 by the Apollo 17 astronauts, the last men to walk on the Moon.

mission to Mars as a positive goal: not as what has been called a 'sprint' mission, but to establish outposts that would be followed by permanent bases. (The mission plan that was most highly favoured is outlined on pages 46–9.)

The Case for Mars

Before the Space Commission's report, talk of manned missions to Mars had been virtually taboo within NASA. But outside the Agency, the idea had been kept alive. In 1981 a small group, including some NASA employees but with no official backing, had organized a conference at the University of Colorado to discuss ways of going to Mars. Calling itself 'The Case for Mars', the conference saw the start of the 'Mars Underground'.

In 1984 the group held its second, much larger, conference. And by the summer of 1987, when the third was held, the Mars enthusiasts felt they had achieved their goals. A NASA-backed study of manned missions had already been published in May 1986, based on a year-long study by some 140 scientists and engineers. A hefty, two-volume report numbering over a thousand pages of text was the fruit of their labours. It covered manned Mars missions in unprecedented detail, from crew selection to the design of spacesuits for use on the Martian surface. Mars was no longer taboo: it was being actively proposed as a goal to reinvigorate the US space programme in the wake of the *Challenger* accident (see page 51).

The third 'Case for Mars' conference opened on 20 July 1987, 18 years to the day since Apollo 11 landed on the Moon. For the very first time, delegates were addressed by the incumbent NASA Administrator. The message, Dr. James Fletcher said, was clear: Americans *would* one day go to Mars. But the question 'When? he left unanswered.

Will history repeat itself?

The 1960s were dominated by the race to the Moon: 20 years later, the superpowers are on the threshold of a race to Mars. In the Moon race the United States at first lagged badly, but triumphed in the end. In the Mars race America is again trailing; can it take the lead again?

The space age opened at the height of the Cold War between the Soviet Union and the United States. It was inevitable that the ideological battle between communism and capitalism should be carried into space. Largely at the behest of Nikita Khrushchev, who saw achievements in space as excellent propaganda, the Soviets pulled off the first space spectaculars—the first satellite, the first animal in space, the first probe to reach the Moon. . .

In 1958, the year after Sputnik 1, the fledgling American space effort was put into the hands of a new government agency—NASA, the National Aeronautics and Space Administration. Its first major goal was Project Mercury—sending men into orbit. The first steps towards this were short 'hops' into space. Commander Alan B Shepard was scheduled to fly on the first of these suborbital flights in May 1961.

Throughout early 1961 there were clear signs that the Soviet Union planned to send men into orbit. Throughout the winter and early spring, the Russians attempted a series of tests with animals on board. On 9 March, Sputnik 4 was returned safely from orbit with a dog as its passenger. A day later, another dog was recovered safely.

On 22 March, President John F Kennedy met with senior advisers to discuss space and, more specifically, how to close the gap with the Soviets. He was looking for a way to galvanize the US space effort, and charged Vice-President Lyndon B Johnson to look into

the matter. On 28 April Johnson reported back: 'To reach the Moon is a risk, but a risk we must take. . . . [In] the eyes of the world, first in space is first, period. Second in space is second in everything.'

Kennedy's advisers were divided. Some felt that the lunar adventure was too risky and too costly: Science Adviser Jerome Wiesner felt that it would eat up all the budgets for scientific research.

Then, on 12 April 1961, the first human being flew in space. Major Yuri Alekseyevich Gagarin made one orbit of the Earth in his Vostok 1 capsule and returned safely. This son of a textile worker won an assured place in history, and his country's highest accolade—Hero of the Soviet Union.

Alan Shepard's suborbital 'lob' went ahead the following month, on 15 May, and was televised live.

Yuri Gagarin in the Vostok 1 capsule, shortly before launch in April 1961.

His 15 minutes and 22 seconds of weightlessness could not compare with Gagarin's one-orbit flight, but it excited the American public. Kennedy made up his mind.

On 25 May 1961 he addressed Congress and unveiled his grand scheme: 'This nation should commit itself to achieving the goal, before the decade is out, of landing a man on the Moon and returning him safely to Earth. No single space project will be more impressive to mankind or more important for the long-range exploration of space; and none will be so difficult to accomplish.' This was the birth of the Apollo project.

Considering the magnitude of the undertaking, it evoked little response in the American press. Kennedy's plan was dismissed as 'Moon madness' and political issues such as the abortive Bay of Pigs invasion in Cuba received greater attention. And yet Kennedy's intention that American science and technology would be whipped up into a frenzy of activity was realized.

It is difficult to decide which is more incredible: Kennedy's go-ahead for the task of sending men to the Moon within the decade, or the fact of its accomplishment. In 1961 NASA launched just two astronauts on suborbital 'hops', yet it was already looking at spacecraft rendezvous in lunar orbit. The following year, on 20 February, John Glenn became the first American to orbit the Earth. He was followed by three others at various times over the next 15 months. NASA's next step was to build the two-man Gemini spacecraft, testing major flight components for Apollo and practising manoeuvring and docking in Earth orbit.

In eight years NASA developed the technology, experience and infrastructure to do exactly as Kennedy had decreed. By that massive research and development effort the USA won the race to the Moon.

Though Soviet authorities later denied it, they had their own manned lunar programme. There are many clues to this: cosmonauts referred to it, the later unmanned Zond flights were dress rehearsals, and the Soyuz orbital stage was designed for the Moon's gravity field. But the Soviet programme suffered many failures. The final nail in the coffin was the failure of the original version of the Heavy Lift Vehicle during the summer of 1969 (see page 42).

During the 1970s the Soviets turned their attention to building and operating space stations in Earth orbit. The experience they gained with seven Salyut stations, and then with the permanently crewed Mir, is invaluable for manned exploration of Mars.

For NASA it is ironic that victory in the Moon race was rewarded with indifference from the American public. But the Apollo experience showed that once Congressional approval is given, America can achieve major national goals remarkably quickly. To compete in the Mars race, the United States needs only to summon the political will.

Above: legacy of Apollo—a footprint in the pristine lunar soil. Top: The Apollo 11 crew were (left to right) Neil Armstrong, Mike Collins, Buzz Aldrin. Right: Jim Irwin, Apollo 15 astronaut.

For NASA's Viking programme scientists, exulting after the brilliant success of their mission in 1976 (see page 17), the next step in Mars exploration was obvious: the return of soil samples to Earth. During 1977 a committee known as the Mars Science Working Group convened to consider how this could be done. They suggested that a follow-on to Viking should be undertaken as early as 1984. After a decade in which NASA's planetary exploration programme dwindled, that mission was still 10 years away— and still by no means certain to take place at all. The only project definitely on the cards was a small orbiter, much less sophisticated than the Vikings, known simply as the Mars Observer. Since it is due to be launched by the Shuttle, the *Challenger* disaster has

affected this project like so many others: its launch date has slipped back two years, to 1992.

Planetary science has always been the Cinderella of space research. NASA has launched no new probes to the planets since the Pioneer Venus craft of 1978. The delays and cost overruns in the development of the Space Shuttle began to eat into NASA's budget. Very soon the cupboard was bare. And President Reagan's freeze on Federal spending in the early 1980s curtailed any ambitious plans for the future. The message was clear enough: there was no room for large-scale, expensive missions.

Partly to appease the enraged planetary science community, but also with the need for cuts in mind, NASA formed a large Solar System Exploration Committee in 1982 to formulate new, more

modest plans. The following year its deliberations were published: a new generation of spacecraft, to be built in production-line fashion using existing technology, should form the basis for further planetary exploration. The Committee's first mission was given the title of Mars Geochemical/Climatology Orbiter and to date is the only project to have been taken up. In early 1985, when its funding was approved, its name was changed to the Mars Observer.

An orbit made for mapping

The Mars Observer will use a new booster, the Transfer Orbit Stage (TOS), to head towards Mars. After a journey of 353 days it will enter a highly elliptical orbit around Mars. Over the next three months, the orbit will be changed to a low, circular polar orbit,

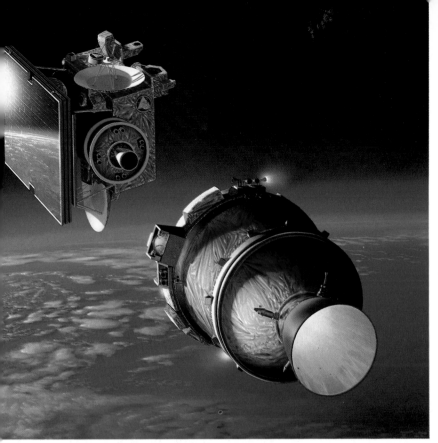

Above: An artist's impression of the Mars Observer separating from the Transfer Orbit Stage (TOS). The TOS fires for just over two minutes to place the spacecraft on the correct transfer orbit for Mars.

Left: The Transfer Orbit Stage with scientific payload deployed from the Space Shuttle.

almost exactly at right angles to the equator. This 'mapping orbit' will allow the spacecraft to observe the whole of the planet in the course of time. The nominal duration of the Mars Observer mission is one Martian year (687 days).

The choice of orbit is crucial for the Mars Observer's scientific objectives. The spacecraft will circle the planet from pole to pole at an altitude of 224 miles (361 kilometres), taking 117 minutes for one revolution. The craft will travel southward on the daylight side of the planet, and northward on the night side. The orbit will be Sun-synchronous with respect to the spacecraft: in other words, as viewed from the Mars Observer, the Sun will be at roughly the same position in the sky whenever the spacecraft crosses the equator. Thus when it crosses the equator in daylight the local time will be 2 p.m. This position of the Sun is ideal for mapping, since shadows are the right length to reveal

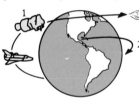

Mars Observer mission profile

1. Mars Observer and TOS deployed from space shuttle.

Launch date: 14 Sept. 1992.

2. TOS fires for 144 seconds to attain Earth escape velocity.

3. 23 minutes later, TOS and Mars Observer separate.

4. Mars Observer in 'cruise' phase. Deploys high gain antenna for communication with Earth. In flight calibration of scientific instruments.

5. After journey of 353 days, Mars Observer fires engines for 19 minutes to reach Mars orbit.

Arrival date: 2 Sept. 1993.

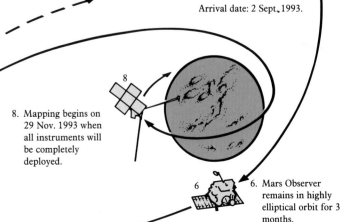

7. After 'drifting' to required orbit, motors fired to enter mapping orbit.

8. Mapping begins on 29 Nov. 1993 when all instruments will be completely deployed.

6. Mars Observer remains in highly elliptical orbit for 3 months. Instruments tested and calibrated.

Mars Observer—scientific payload

Payload mass: *280lb (127kg)*
Payload power: *148.5 watts*
Data rate at Mars: *64 kilobits/sec*

Scientific instruments

GAMMA RAY SPECTROMETER (GRS)

This instrument has a high spectral resolution at gamma-ray wavelengths and is mounted at the end of a 20-foot (six-metre) boom. It will determine the abundance of elements such as potassium, iron, uranium, silicon and thorium, among others, over the planet.

MAGNETOMETER (MAG)

Another boom-mounted instrument, it will measure the strength of the Martian magnetic field, poorly understood at present.

MARS OBSERVER CAMERA (MOC)

A line-scanning television camera with both wide- and narrow-angle fields of view. The camera will look at Mars at low resolution to monitor the planet's weather. At selected sites it will return pictures with resolution down to 4.5 feet (1.4 metres) on the surface.

PRESSURE MODULATOR INFRARED RADIOMETER (PMIRR)

A scanning device used to 'profile' the atmospheric structure at infrared wavelengths. It will monitor changes in temperature, dust and water vapour with latitude and longitude as well as season.

RADAR ALTIMETER (RAR)

This instrument will 'bounce' 13.6 GHz radio waves off the surface: the radar 'echo' will give a detailed picture of surface topography.

RADIO SCIENCE (RS)

As part of the telecommunications link, an ultrastable oscillator will generate a signal with a very precise frequency. Minor fluctuations in signal strength reflect gravitational anomalies, giving clues to Mars's internal structure.

THERMAL EMISSION SPECTROMETER (TES)

This instrument will observe the thermal emissions of rocks on the surface, frost and clouds. Their chemical composition can be worked out as every element has a unique spectral 'signature'.

VISUAL AND INFRARED MAPPING SPECTROMETER (VIMS)

The VIMS observes the chemical composition of the surface across 320 spectral channels from visible wavelengths to infrared ones. It will map the mineral and chemical composition of the surface rocks. It will be able to detect water ice in crater floors and the amount of carbon dioxide in the atmosphere.

ground relief well.

Once the mission has been completed, the orbit's height will be increased to ensure that the spacecraft does not hit the surface before the year 2038. This is to comply with international quarantine regulations.

The Mars Observer shares the same basic design as RCA's Satcom K communications satellite and has electronic subsystems used on the TIROS and Defense Meterological Satellite Program (DMSP) weather satellites. In this way development costs have been minimized. The spacecraft as a whole is 'nadir-oriented'—its instruments always point downwards towards the planet, avoiding the complexities of mounting them on a scan platform.

The spacecraft has a full complement of eight scientific instruments (see box). Its major scientific aims are to investigate the planet's meteorology, climate and atmospheric chemistry, and the composition of the whole of the Martian surface. An instrument such as the VIMS (Visual and Infrared Mapping Spectrometer—see box), which has a field of view covering an area 25 miles (40 kilometres) across at the surface, will be able to survey the whole of the planet in about 50 days. Over the Martian year the Mars Observer will be able to go through this cycle of observation about 12 times. By covering the planet in this way, the Mars Observer will be able to see seasonal changes, which is particularly useful for monitoring weather patterns.

Sampling Mars

Though not originally designed for such a purpose, the Mars Observer may be modified to help in the selection of landing sites for the next American venture to the red planet: an unmanned rover and sample return mission. NASA is considering a number of options

for a mission that could be mounted before the end of the century. At present they are conceptual studies; if approved, the funding of such a mission would begin in earnest in the 1993 budget.

Though the details are not fixed, the overall concept is clear: a lander will separate from an orbiting spacecraft, and after touchdown will despatch a rover to travel across the surface. After exploring for about a year, during which time it would selectively sample the surface rocks, the roving vehicle would return to the lander. The samples would be transferred to a small ascent stage, which would rendezvous with the orbiter. The sample container would then transfer to an Earth return vehicle forming part of the orbiter. Current plans envisage that the samples would be analysed aboard the US Space Station to prevent contamination of the Earth.

Artist's impression of the US rover and sample return craft (above), and a rover variant (below).

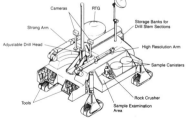

Within this general mission profile, there are a number of factors that are under consideration. For example, will the spacecraft systems require new technology to be developed, particularly for the automated roving activities? Options range from local sampling

only (with only a small tethered device, or possibly no rover at all) to a rover capable of traversing many hundreds of miles.

Because of the time lag in communications between Earth and Mars, operations cannot be in 'real time'—that is, with direct control from the Earth. So the rover will have to be a robot, with a high degree of autonomy, capable of making its own 'decisions'. But the degree of automation depends on the amount of scientific payload that the rover can carry. The current design studies estimate 880 to 3300 pounds (400 to 1500 kilograms) of scientific payload. This weight can be increased if the amount of rocket fuel required during the mission can be reduced.

Reducing fuel cost

One way of reducing it is to use aerobraking techniques during the spacecraft's insertion into orbit arounds Mars (see page 81). Perhaps it would help, too, if the various elements of the mission (say, orbiter, lander, Earth return stage) were flown separately—a possibility that has not been ruled out.

Mars roving involves many difficult operations, including traversing terrain whose nature is unknown, finding geologically interesting rock samples, selecting and identifying them, and storing them for long periods. Navigating on the Martian surface involves identifying possible local routes and planning journeys, as well as knowledge of the rover's global position. These problems will also face astronauts on the Martian surface, and will be considered in greater detail in Part 5.

The Jet Propulsion Laboratory and the Johnson Space Center in Houston are currently assessing the various mission options. By the middle of 1989 they will have decided which are most suitable and will proceed with more detailed analyses. The final mission profile will be worked out by 1993, for a launch in the 1998 launch window.

SOVIET MISSION PROFILE

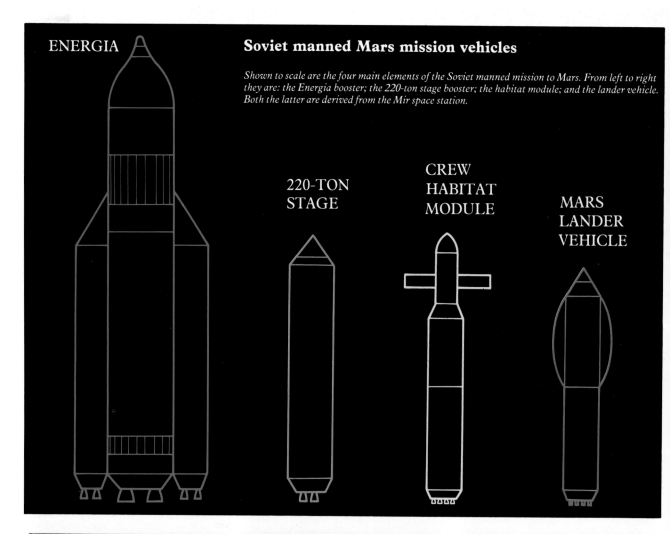

ENERGIA

Soviet manned Mars mission vehicles

Shown to scale are the four main elements of the Soviet manned mission to Mars. From left to right they are: the Energia booster; the 220-ton stage booster; the habitat module; and the lander vehicle. Both the latter are derived from the Mir space station.

220-TON
STAGE

CREW
HABITAT
MODULE

MARS
LANDER
VEHICLE

1 The first Energia lifts the combination of the unmanned lander and two Proton stages to orbit.

2 The Energia's six strap-on boosters are jettisoned and the core carries the payload to low Earth orbit.

3 The lander, with fuelled Proton stages attached, orbits the Earth.

4 The second Energia lifts a 220-ton LOX/H₂ rocket into orbit.

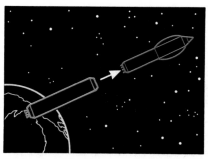

5 Once in orbit, this stage automatically docks with the lander assembly.

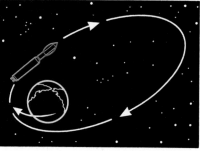

6 The first-stage rocket is fired to put the lander into an elliptical Earth-synchronous orbit.

7 The spent rocket is jettisoned, and the Proton 3 stage is fired a day later to launch the assembly towards Mars.

8 The third Energia lifts the Mir-based habitat module, attached to a fully fuelled Proton second stage, into parking orbit.

9 The crew are launched from Earth in a conventional Soyuz vehicle carried by an SL-4 booster.

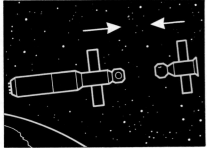

10 The Soyuz joins the habitat module assembly and is linked with the end docking port.

11 The crew transfer to the habitat module, their home for the journeys to and from Mars.

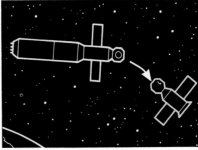

12 Once inside the habitat module, the crew seal the hatches and jettison Soyuz.

13 The fourth Energia lifts another 220-ton stage into low Earth orbit.

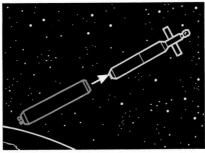

14 The 220-ton stage docks with the habitat module assembly.

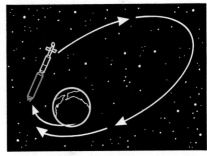

15 The 220-ton stage is fired to put the crew transit assembly into Earth-synchronous parking orbit.

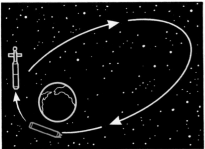

16 After jettisoning the spent 220-ton stage, the Proton stage fires to launch the habitat module towards Mars.

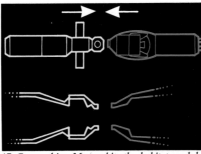

17 On reaching Mars orbit, the habitat module assembly is linked to the lander assembly, already in orbit.

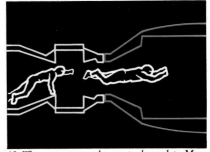

18 The cosmonauts who are to descend to Mars transfer to the lander.

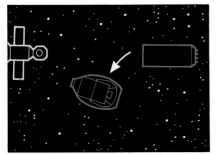

19 The lander assembly separates, and the Proton second stage fires to brake it before being jettisoned.

20 The lander's aeroshell is deployed and the descent begins.

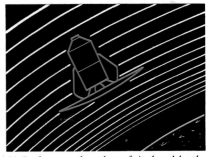

21 In the atmosphere the craft is slowed by the drag on its aeroshell.

Although there are numerous ways in which a mission to Mars could be organized, the mission profile outlined here is, according to Western expert opinion, the one most likely to be adopted by the USSR.

The assult on Mars is conducted by a crew of at least three men in *two* spaceships. At Tyuratam the Soviets have three launch pads for their new Energia boosters. They will need four Energias, each with six strap-on boosters, to lift the components of their Mars fleet into orbit around the Earth. There they are assembled to form two fully fuelled multi-stage rockets. It is less costly in fuel to launch these vehicles from orbit, weightless and unhampered by air resistance, than from Earth.

The first Energia lifts an unmanned payload consisting of the Mars lander attached to fully fuelled Proton second and third stages—a combination weighing about 220 tons.

The second Energia puts up a first-stage rocket, fully fuelled with liquid oxygen and liquid hydrogen (LOX/H$_2$). This automatically docks with the lander combination already in orbit. The assembly then goes into a 'two-impulse escape trajectory'. First the LOX/H$_2$ stage is fired to put the assembly into a highly elliptical 'Earth-synchronous' parking orbit that takes the craft as high as 48,250 miles (77,700 kilometres) and as low as 4100 miles (600 kilometres), measured from the centre of the Earth. An Earth-synchronous orbit is one that the spacecraft traverses in exactly the time it takes the Earth to turn once on its axis—24 hours, less four minutes.

Triggered from the ground

This means that when the spacecraft returns to the lowest point of its orbit, it will be exactly over the launch site again: important when it is required to activate something on board—in this case, the firing of an engine—by a radio signal from the ground.

So, as the orbiting assembly passes over Tyuratam, one day after launch, ground control puts it on course for Mars by firing its Proton 2 stage (the first stage, its fuel exhausted, has been jettisoned).

With the craft carrying the Mars lander safely on its way, the man-carrying craft can now be assembled in Earth orbit. Another Energia lifts a fully fuelled Proton second stage attached to the habitat module, to which is attached a Soyuz Earth return capsule. This payload is again in the region of 220 tons.

The fourth Energia puts up a first stage, fully fuelled with liquid oxygen and liquid hydrogen. It automatically docks with the habitat combination.

When docking is complete, cosmonauts can be sent up in a conventional Soyuz craft or its successor, possibly a small spaceplane.

After a thorough checkout, the manned assembly goes through the same sequence of events as the first one: first it enters an Earth-synchronous orbit and then, a day after launch, its Proton 2 stage is fired, and the cosmonauts begin the voyage to Mars.

Arrival at Mars

So now there are two craft heading for Mars: the first, comprising Proton second and third stages linked to the, as yet unmanned, Mars lander; the second, a Proton second stage attached to the transit vehicle in which the cosmonauts make the journey.

It is possible that the Mars fleet may be even larger than this. Unmanned cargo vessels carrying supplies might be sent ahead to go into Mars orbit and await the arrival of the cosmonauts.

When the manned ship approaches Mars nine months later, it finds the vehicle carrying the lander already in its parking orbit. The cosmonauts use their Proton second-stage engine as a retro-rocket to slow down and go into Mars orbit. They dock with the lander assembly, and at least two crew members crawl into it, undock and prepare to descend to the surface. The rest of the crew will remain in orbit while their colleagues explore the surface.

A short burn by the Proton second stage slows the lander assembly and it starts to descend. The Proton is jettisoned and the lander enters the Martian atmosphere. It is braked further by atmospheric drag on the 'aeroshell' on which it is mounted. When the speed has dropped to 2600 feet per second (800 metres per second), the shield is jettisoned and parachutes are deployed. Close to the surface, with the speed slowed to 330 feet per second (100 metres per second), the Proton third-stage engine fires for a cushioned landing.

The journey home

The exploration and study of the surface could last from a few months to a year. It must however, be brought to a close when the positions of the Earth and Mars are favourable for the return journey. The cosmonauts leave the Martian surface by firing the Proton 3 engine on the lander. Once in orbit around Mars, they dock with the still-orbiting habitat module and transfer to it. Then they jettison the lander. The fuel that remains in the Proton 2 attached to the habitat module is used for a 'trans-Earth injection burn', placing the vehicle on the trajectory that will take it to the vicinity of the Earth.

Nine months later the cosmonauts approach the Earth. Probably they will transfer into the Soyuz return module about an hour before their touchdown. They separate from the habitat module, which burns up in the Earth's atmosphere, and come in as the Apollo astronauts used to do, 'skipping' on the upper layers of the atmosphere to slow down before deploying parachutes.

22 When the speed has been reduced to 2600ft/sec (800m/sec), the aeroshell is jettisoned.

23 Parachutes slow the lander to about 330ft/sec (100m/sec).

24 The parachutes are jettisoned and the lander legs are deployed.

25 The Proton third stage fires to slow the lander's descent still more.

26 A final burst from the rocket cushions the touchdown on the surface.

27 With the lander on the surface, the crew climb into pressure suits ready for their first 'Mars walk'.

28 The lander hatch is opened and the crew pass through the airlock and climb down the ladder to the surface.

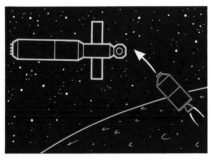

29 When it is time to depart for the Earth, the lander's Proton rocket lifts the crew into orbit.

30 The lander docks with the orbiting habitat module and the exploration crew transfer back into it.

31 The lander is jettisoned and the crew set course for the return to Earth.

32 Nearing the earth, the crew enter the Soyuz Earth return capsule and jettison the spent Proton third stage.

33 The return capsule separates and its braking rockets are fired.

34 During re-entry the capsule rotates so that its heat shield is forward.

35 The capsule 'skips' off the upper layers of the atmosphere and the heat shield glows white-hot.

36 The final part of the descent is by parachute, ending with a rocket-cushioned touchdown.

SOVIET CRAFT
DESIGN AND ASSEMBLY

Energia and payload in launch configuration

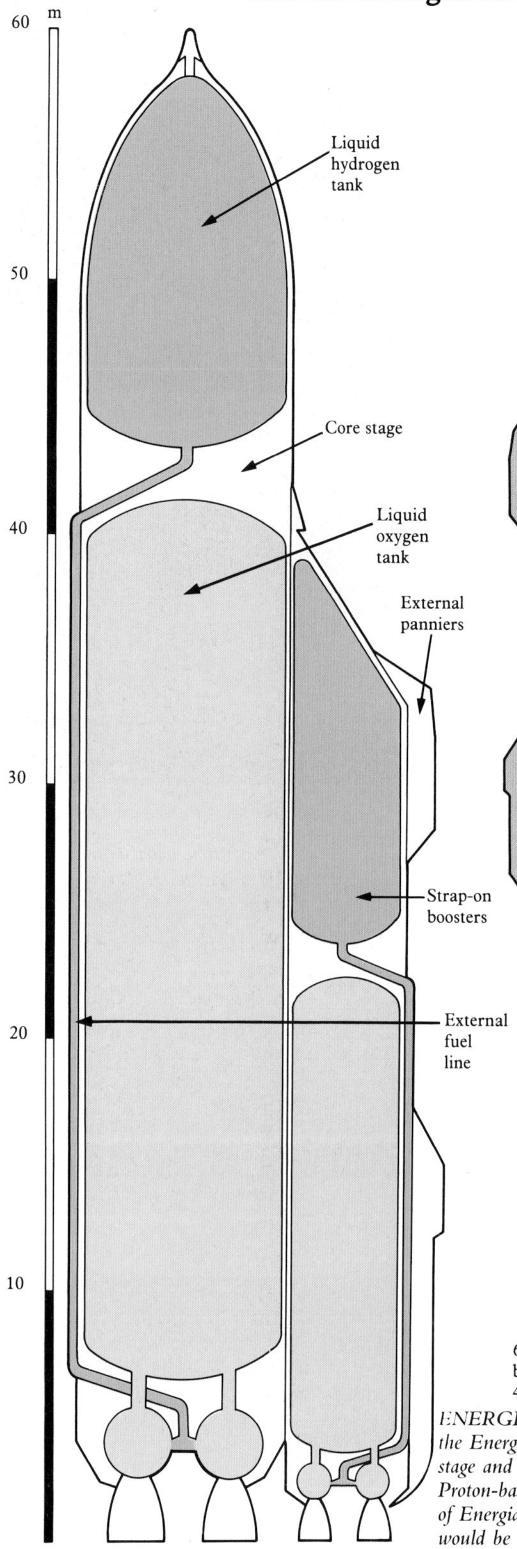

Liquid hydrogen tank

Core stage

Liquid oxygen tank

External panniers

Strap-on boosters

External fuel line

Core 8m dia.

Payload 4.15m dia

6 Strap-on boosters 4.3m dia.

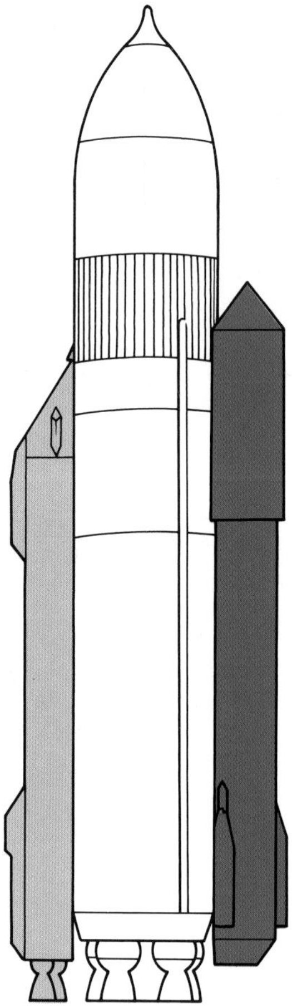

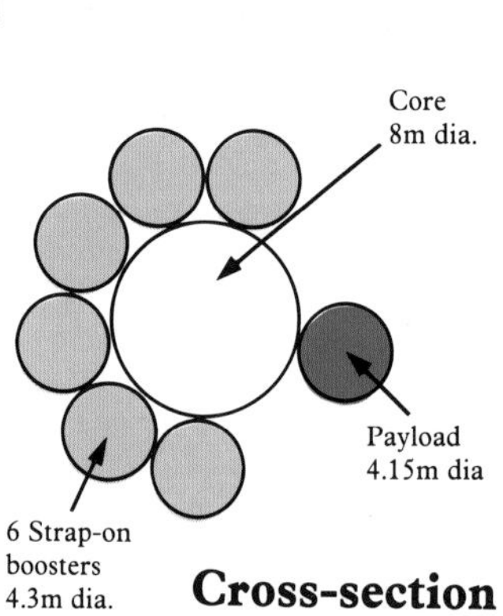

Cross-section

ENERGIA ROCKET. A cutaway through the Energia booster (left) details the main core stage and one of the strap-on boosters. A Proton-based vehicle can be mounted on the side of Energia (top). For the Mars trip the assembly would be configured with six strap-ons as shown in the cross-section (above).

Energia

The Soviet Mars ships will possess a pedigree going back to the days of the Moon race in the late sixties. At that time the USSR built a giant Moon rocket, known variously as the G-1, the SL-15 'Super Booster', the TT-5, or 'Webb's Giant' (after James Webb, the NASA Administrator who had first revealed its existence to the West).

This vehicle was a shade smaller than the American Saturn V—325 feet (100 metres) tall as opposed to Saturn's 363 feet (111 metres)—but it was designed to have greater thrust. But the Soviets got into deep trouble with this launcher. In June 1969, a few weeks before Neil Armstrong stepped onto the Moon, the Russian rocket was on its launch pad being fuelled when a leak in the second stage led to an explosion that wrecked the entire launch area. American satellite pictures taken at the time, still classified, are said to show the debris.

It was more than two years before a second attempt was made to launch the Super Booster. Again there was disaster. It had reached a height of only 7.5 miles (12 kilometres) when heavy vibrations caused it to shake itself to pieces. There was a third failure in November 1972, when the rocket had to be destroyed soon after launch.

But its importance to long-term Soviet space plans meant the rocket could not be entirely abandoned, and it was decided to redesign it. The whole thing was done over the next 15 years in great secrecy, although the new rocket was glimpsed by Western spy satellites in 1983. In the West it was codenamed 'HLV', for 'heavy lift vehicle'. The Russians refused to say anything about it, even after the flight-readiness firing of its engines in early 1987.

The Energia rocket

Then, in May 1987, the Soviets test-launched the new rocket and it was a qualified success. They called the new vehicle 'Energia'.

After this launch, the Russians not only told the world at once but released quite detailed television pictures of the rocket's structure within a few hours. By contrast, the first time the Soviets showed television pictures of the Proton booster was 20 years after it had come into operational use.

Energia comprises a central core 197 feet (60 metres) tall and 26 feet (8 metres) wide, surrounded by strap-on boosters, each 131 feet (40 metres) tall and 14 feet (4.3 metres) wide. On its first launch there were four of these. The core contains four high-performance engines fuelled by 1000 tons of hydrogen/oxygen propellant. The booster rockets use kerosene and oxygen.

The boosters are set in pairs. If the payload is mounted at the side of the core, as it was on the test flight, there can be up to six boosters. They and the central core are intended to be fully recoverable, though they were not in fact recovered on the initial flight.

Using four strap-ons Energia will be able to lift well over 110 tons of payload; in fact, fully fuelled (which it was not on its maiden flight), it is believed that it could lift 170 tons. With six strap-ons it will be able to lift 230 tons of payload into low Earth orbit.

This is said by the Soviets to be Energia's maximum capability.

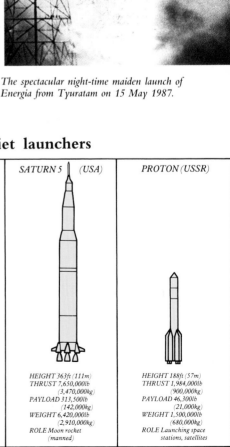

The spectacular night-time maiden launch of Energia from Tyuratam on 15 May 1987.

Proton-based vehicles

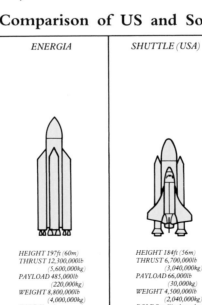

Habitat module

Mars lander

Lander legs stowed away

Aeroshell folded away

The habitat module in which the crew will travel to and from Mars is essentially a modified Mir space station. A Soyuz Earth return capsule is mounted at the bottom.

The Mir-based habitat module (far left) and Mars lander vehicle (left) mounted on modified Proton 2 launch vehicles.

The Mars lander comprises a Mir-like pressurized cabin with a Proton 3 rocket engine. The lander legs and aerobrake are shown in the stowed-away position.

Comparison of US and Soviet launchers

ENERGIA	SHUTTLE (USA)	SATURN 5 (USA)	PROTON (USSR)
HEIGHT 197ft (60m)	HEIGHT 184ft (56m)	HEIGHT 363ft (111m)	HEIGHT 188ft (57m)
THRUST 12,300,000lb (5,600,000kg)	THRUST 6,700,000lb (3,040,000kg)	THRUST 7,650,000lb (3,470,000kg)	THRUST 1,984,000lb (900,000kg)
PAYLOAD 485,000lb (220,000kg)	PAYLOAD 66,000lb (30,000kg)	PAYLOAD 313,500lb (142,000kg)	PAYLOAD 46,300lb (21,000kg)
WEIGHT 8,800,000lb (4,000,000kg)	WEIGHT 4,500,000lb (2,040,000kg)	WEIGHT 6,420,000lb (2,910,000kg)	WEIGHT 1,500,000lb (680,000kg)
ROLE Shuttle, manned planet missions	ROLE Satellite launcher (manned)	ROLE Moon rocket (manned)	ROLE Launching space stations, satellites

The Mars Lander

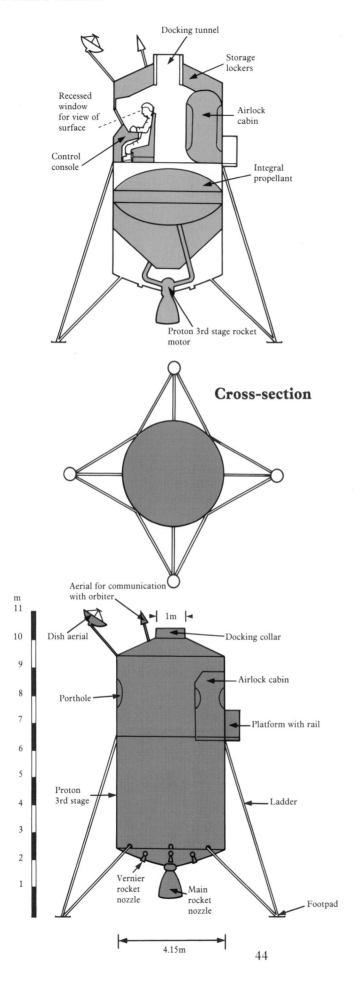

Docking tunnel

Storage lockers

Recessed window for view of surface

Airlock cabin

Control console

Integral propellant

Proton 3rd stage rocket motor

Cross-section

Aerial for communication with orbiter

1m

Dish aerial

Docking collar

Porthole

Airlock cabin

Platform with rail

Proton 3rd stage

Ladder

Vernier rocket nozzle

Main rocket nozzle

Footpad

4.15m

But if a version is used in which the payload is placed on top of the core, the number of strap-ons could be increased to eight, and this would mean that much heavier payloads could be carried.

Compare these weights with the five tons carried by the European Ariane launcher, the 30 tons that can be carried in the US Space Shuttle's cargo bay, or even Saturn V's 150 tons.

The Proton stages

The upper stages of the Proton rocket have been used regularly since the 1960s to launch satellites. They were 'man-rated' as part of the Soviet Moon programme. (Man-rating is the adaptation of hardware to the requirements of manned spaceflight: thus rocket engines must be free from excessive vibration, must be able to be switched on and off to manoeuvre in space, and must be totally reliable.) Proton rocket stages will play a key role throughout the Mars mission.

The habitat module

Cosmonauts first learned to live for extended periods in space in the Salyut space station. Salyut crews set up endurance records first of six months, then of seven—periods comparable with the journey time to Mars.

The Mir space station, launched in February 1986, was a follow-up to Salyut, designed for even longer flights. The Russians have said that Mir, with its six ports for attaching modules, was designed with Mars in mind. With only slight modifications, it will be used as the habitat module in which the cosmonauts will travel to Mars and back.

The Mars lander

When the USSR lost the race to put a man on the Moon, it did not scrap its Moon lander. It was based on the Soyuz craft (and the unmanned Zond variant of Soyuz actually made several trips to the Moon). Soyuz was adapted to

44

ferry cosmonauts to and from orbit but it retained its original interior, designed for use on the Moon. There is a couch and a worktop that would be horizontal after landing, and a hatch positioned so that cosmonauts could climb down to the Moon's surface by ladder.

While Soyuz could be used to land on the Moon when the Soviets return there, it is believed that a derivative of the Mir space station is the most likely candidate for a Mars lander. Mir could easily be modified to feature a pressurized module through which cosmonauts could pass onto a ladder leading down to the Martian surface. Cosmonauts on their way out of the craft would enter this 'airlock', the pressure would be reduced to match that of the Martian atmosphere and only then would the hatch be opened. When the cosmonauts had re-entered the module and closed the hatch, the pressure would be increased to match that of the lander's interior. This would avoid the necessity of constantly depressurizing and repressurizing the entire lander.

The Heavy Cosmos cargo vessels

A full Mars mission could last nearly three years, and it would be necessary to carry every pound of oxygen, food and water consumed in that time. Even assuming that 50 per cent of the water could be recycled, about 10 pounds (4.5 kilograms) of consumables would be used every day—a total of over four tons.

It is possible that spacecraft known as Modulny, or Heavy Cosmos, roughly the size of the old Salyut, will be used to transport some of these supplies. Four of these mysterious craft have been flown, three in conjunction with Salyut, but the Russians have never revealed their purpose. Such craft could carry supplies into orbit around Mars for use after the arrival of the two main ships.

HABITAT MODULE. The habitat for the crew en route to Mars is derived from the Mir space station, first launched in February 1986. This in turn was developed from the earlier Salyut space stations, in which cosmonauts learned to live for several months at a time in space.

CONTROL SECTION. Communications, observation instruments and Mir system controls are housed here.

LIVING SECTION. Sleeping, sanitation and recreation facilities are grouped in this part of the craft.

DOCKING PORTS. Mir has a total of six docking ports, one at each end and four set radially. These are to enable further modules to be linked to Mir when it is functioning as a space station. Only the front-end port is used on the Mars mission.

Mir-based habitat module

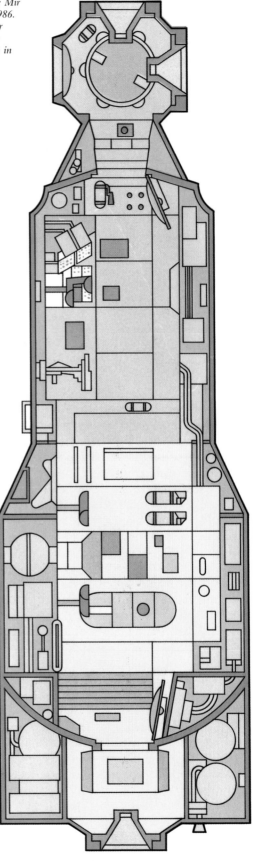

Many ways of going to Mars have been proposed. The most ingenious and the one most favoured by the US National Commission on Space was proposed by Dr John Niehoff of Science Applications International Corporation (SAIC), a technical consultancy based near Chicago. After studying the spatial relationships of the Earth and Mars, he realized that travel between the two planets could be very economical in terms of time and energy.

A cosmic ski-lift

The orbits of Earth and Mars are related to each other in such a way that a spaceship in an appropriate orbit around the Sun could visit both planets at regular intervals. Niehoff plans to exploit this possibility with a Versatile Station for Interplanetary Transport—a name contrived to provide the acronym VISIT.

Niehoff reasons that if four or more interplanetary vehicles were sent into such an orbit, a kind of cosmic ski-lift could be set up, permanently embracing the two planets' orbits and using their gravitational influence to provide much of the necessary power. These travelling space stations would allow astronauts to 'hitch a ride' across the solar system.

The VISITOR vehicles would be assembled in Earth orbit. They would be set spinning at two revolutions per minute to create artificial 'gravity' on board, with a strength about one third of that at the Earth's surface. This is intended to combat the effects of long-term weightlessness on the human body—though some dispute the method's practicality (see page 58). The ViSIT principle is unaffected, however, by the question of whether the travelling space stations will be given spin.

After assembly each space station would be launched into its orbit around the Sun and would begin its regular shuttling between Mars and Earth. When a station swept high above the Earth, two spacecraft would be launched from low Earth orbit to meet it. Each of these craft could carry four crew members, as well as the food and fuel needed for the journey to Mars.

After docking, the astronauts would transfer to habitat modules at the end of the 250-foot (75-metre) arms of the space station. Two people could live in each of the spacerooms, which would be about 36 feet by 16 feet (11 metres by 5 metres), with all the necessary facilities for the nine-month journey to Mars.

Nuclear life support

Other modules on each arm would be designed for use as workshops or laboratories. Even a gymnasium could be included. At the centre of the station would be a 'storm cellar', where astronauts could shelter during solar-flare eruptions to protect themselves from radiation. The necessary energy for life-support systems could be provided by a nuclear power source, with a radiation shield to protect the crew and a black ceramic skirt to radiate waste heat.

On the approach to Mars the crew would transfer to 'biconic' landing vehicles, shaped to allow the use of the Martian atmosphere as a brake before the engines were fired to go into a stable orbit around the planet. At this point booster vehicles shaped like flying saucers would separate from the station and go into a parking orbit around Mars.

The biconics' orbit would be maintained until the crew identified a safe landing place. The crew would descend to the Martian surface, parachutes helping to slow their fall. A final burn of the engines would provide a smooth touchdown.

The length of stay on the surface would depend on the date of arrival of the next travelling space station. The wait could vary from a few months to a year or more. At that time the explorers would get back into the biconics, blast off from the surface and go into a low orbit around Mars. Now they would need the boosters, waiting for them in orbit 180 miles (300 kilometres) high. Each biconic would link up with one of the boosters, which would contain enough fuel to launch the biconic out of orbit for the rendezvous with the interplanetary vehicle. The astronauts would dock with the station and ride back to Earth.

As the mission came to an end the crew would use the biconic spacecraft again, this time to dive towards the Earth. Once more the atmosphere would be used as a brake—this time to put the crew into the same orbit as the American manned space station, which would be operational by then. After docking, a Shuttle would take them back to Earth. The mission would have lasted about three years in all.

Since Niehoff originally proposed his idea, the complicated mathematics used to calculate the VISIT orbits has been further refined. The principle remains the same, but in the latest version the time between successive flybys of Earth and Mars has been minimized. But four of the VISITOR space stations will still need to be put into orbit to provide Mars travellers with an efficient interplanetary 'escalator'.

The elegant VISIT technique would provide a permanent method of regular travel to Mars, essentially geared to long-term exploration of the planet and subsequently to the establishment of a permanent colony.

1 The VISITOR interplanetary vehicle approaches Earth.

2 Two biconic craft are launched from Earth, each carrying four crew members.

3 The biconics dock with the VISITOR craft.

4 The crews transfer from the biconics into the VISITOR vehicle.

5 Crew members head towards the habitat modules of the VISITOR vehicle as it heads off towards Mars.

6 Approaching Mars, the crew enter the biconics and separate from the VISITOR vehicle.

7 Each biconic dips into the Martian atmosphere to decelerate and then enters orbit around Mars.

8 The biconic orbit is finalized once a landing site has been chosen.

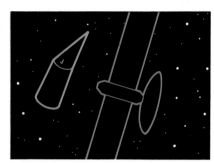

9 The saucer-shaped rocket boosters separate from the VISITOR vehicle.

10 Each booster moves into a parking orbit around Mars.

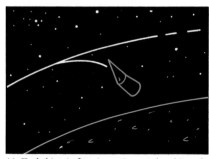

11 Each biconic fires its engines to de-orbit and descend to the surface.

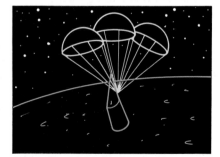

12 Parachutes are deployed to slow the biconics in the Martian atmosphere.

13 Each biconic fires its engines to cushion its touchdown.

14 The crew explores the surface until the next VISITOR vehicle approaches Mars.

15 When it is time to leave, the biconics fire their engines and lift off.

Profile of VISIT orbits

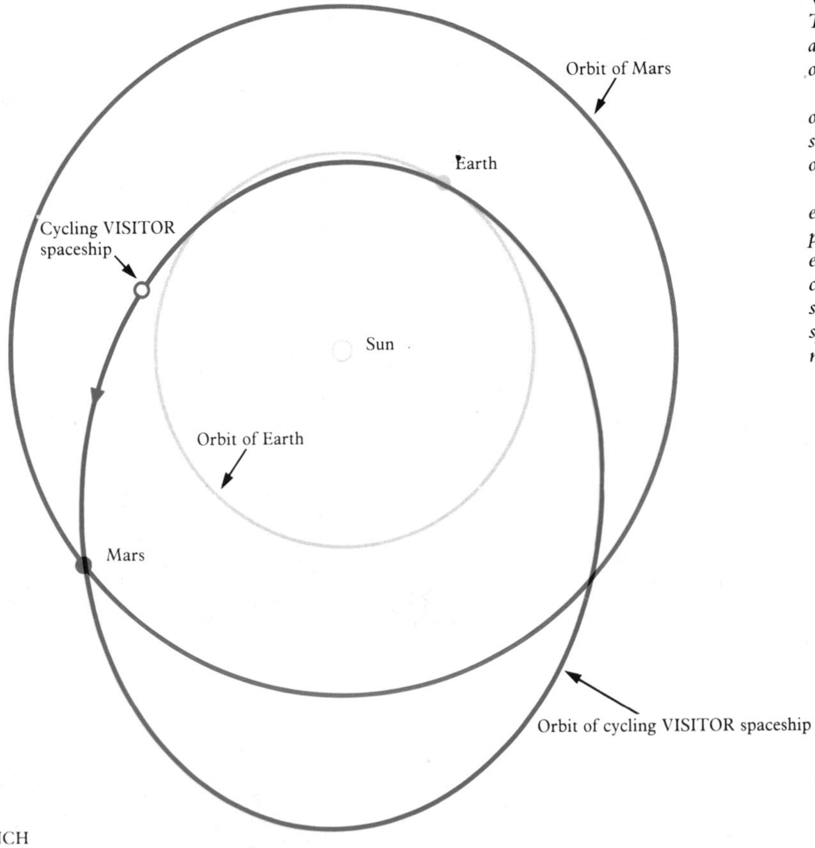

Orbit of Mars

Earth

Cycling VISITOR spaceship

Sun

Orbit of Earth

Mars

Orbit of cycling VISITOR spaceship

VISIT orbit scenario
The orbits of the Earth (blue) and Mars (red) are shown at the left (not to scale). The VISIT orbit (green) neatly crosses between them.

There are many variations on the VISIT orbit theme; below is a simplified scenario showing the progress of a VISITOR spacecraft over a two-year period.

The events shown do not repeat themselves exactly over successive two-year cycles. Over a period of decades the time between successive encounters with either planet will vary considerably. To ensure a reasonably regular supply service, at least four VISITOR spacecraft on slightly different orbits will be needed.

1 LAUNCH
Mission elapsed time (MET) = 0 days

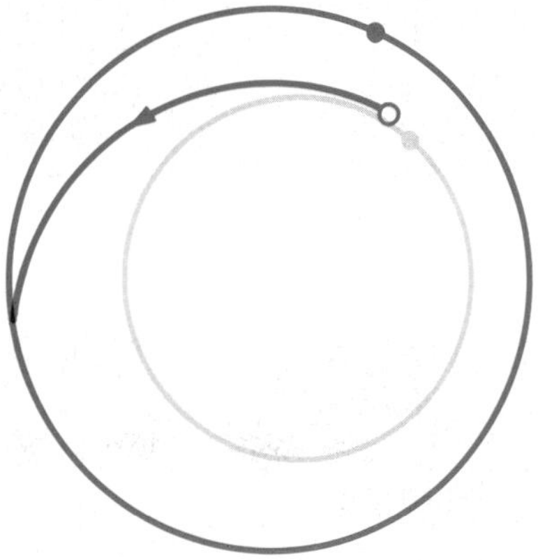

2 FIRST MARS ARRIVAL
MET = 163 days

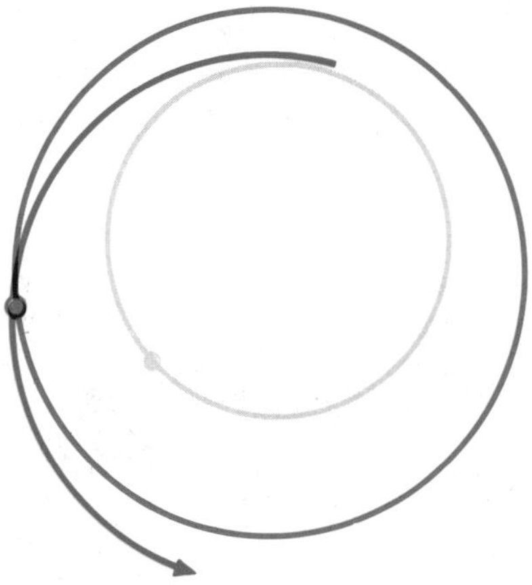

3 MARS-EARTH TRANSIT
MET 365 days

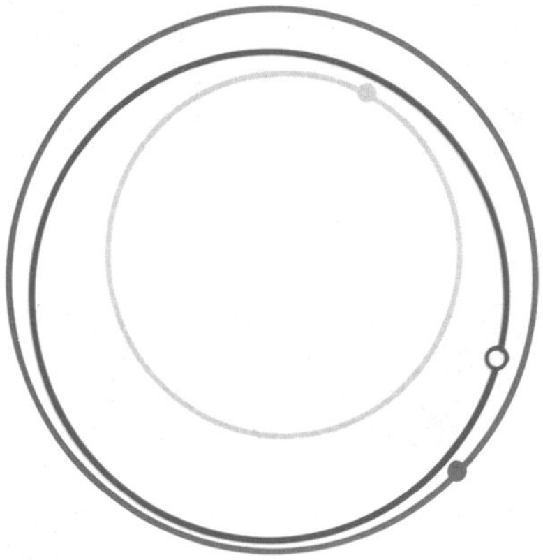

4 EARTH-MARS TRANSIT
MET = 488 days

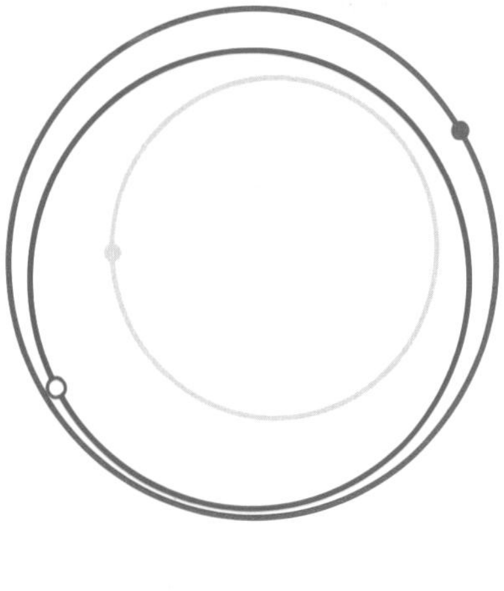

5 RETURN TO EARTH
MET = 611 days

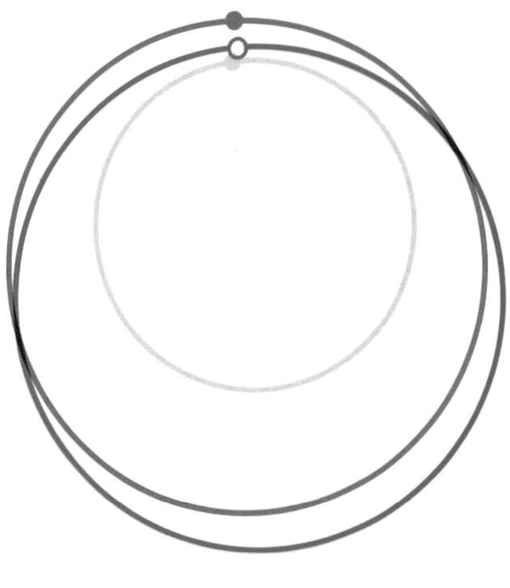

6 SECOND MARS ARRIVAL
MET = 758 days

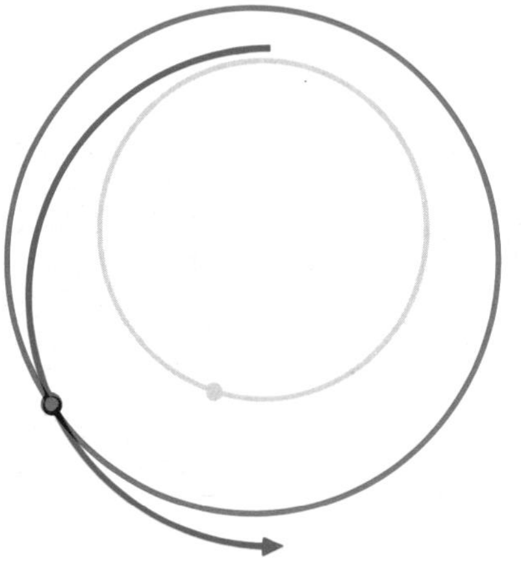

16 Each biconic enters low orbit around Mars.

17 Each biconic heads towards rendezvous with one of the orbiting boosters.

18 The biconic docks with the booster.

19 The booster engine fires to carry the assembly out of Mars orbit and towards the approaching VISITOR vehicle.

20 The combination docks with the VISITOR vehicle.

21 The biconic separates and docks on the other side of the VISITOR vehicle arm.

22 The second biconic and booster repeat the process.

23 The VISITOR vehicle, with the crew safely aboard, heads for Earth.

24 As they near the Earth, the crew return to the biconic craft and undock.

25 They use the Earth's atmosphere to aerobrake and establish Earth orbit.

26 Each biconic finalizes its orbit around the Earth.

27 The biconics dock with the Space Station.

28 The crew members are ferried back to Earth by the Space Shuttle.

SPRINTING TO MARS

Cargo operations

Crew operations

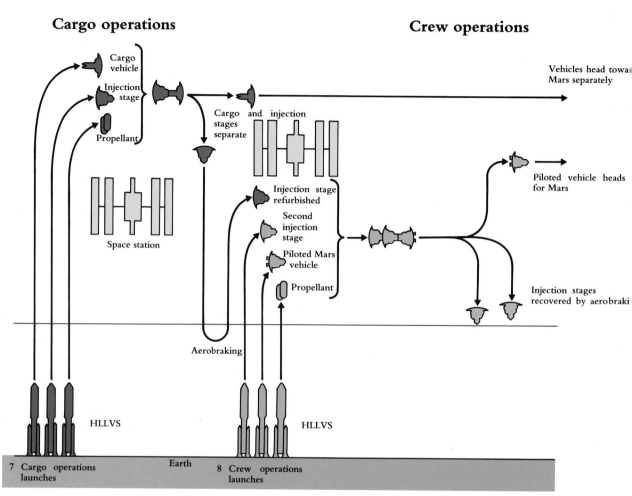

Cargo vehicle

Injection stage

Propellant

Space station

Cargo and injection stages separate

Injection stage refurbished

Second injection stage

Piloted Mars vehicle

Propellant

Aerobraking

HLLVS

HLLVS

7 Cargo operations launches

Earth

8 Crew operations launches

Vehicles head toward Mars separately

Piloted vehicle heads for Mars

Injection stages recovered by aerobraking

Mars orbital/surface operations

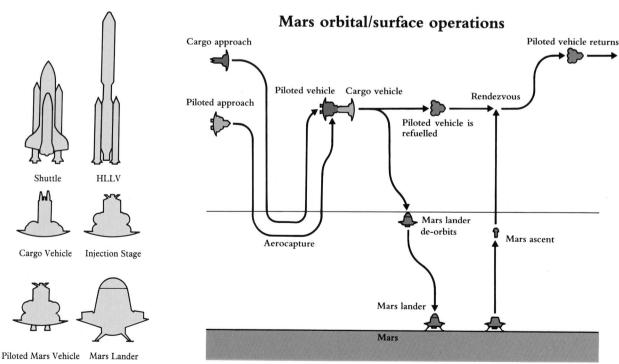

Shuttle

HLLV

Cargo Vehicle

Injection Stage

Piloted Mars Vehicle

Mars Lander

Cargo approach

Piloted approach

Aerocapture

Piloted vehicle   Cargo vehicle

Piloted vehicle is refuelled

Mars lander de-orbits

Mars lander

Rendezvous

Piloted vehicle returns

Mars ascent

Mars

NASA's next steps in space will be the resumption of Shuttle flights and the construction and operation of the Space Station by the middle of the 1990s. So where does Mars fit into the American programme?

A number of expert panels, conscious of NASA's decline, have proposed different ways of invigorating it. In March 1987 the NASA Advisory Committee reported that: 'We should make exploring and prospecting on Mars our primary goal, and should so state publicly.'

NASA responded by forming a new Office of Exploration in June 1987, to look into 'expanding the human presence beyond Earth.' Dr Sally Ride, America's first woman in space, became the Office's first interim director and was given the task of defining new space goals. Her report, *Leadership and America's Future in Space,* was published in August 1987. It detailed four possible initiatives: greater global mapping and coverage by remote imaging satellites;

further exploration of the solar system; setting up an outpost on the Moon; and sending human beings to Mars.

The last two options have aroused most interest. But which is it to be: the Moon or Mars? Reading between the lines, the Ride report suggests that NASA's long-term goal should be continual human exploration of space. The report also suggests that it would be rational to develop technology that could be adapted for either the Moon or Mars.

The main advocate of this latest 'Humans to Mars' initiative was John Niehoff, who conceived the VISIT scenario. As opposed to that highly complex plan, his latest ideas envisage 'sprint' missions that would permit the first US astronauts to land on Mars as early as 2005. They would spend two to three weeks there before returning to Earth after a year-long round trip. Such short flight times would be possible because the vehicles used would be much simpler and less massive than Mars ships of

conventional design. Hence they could be accelerated to higher journey speeds at an acceptable cost in fuel.

A sprint mission is outlined here. A heavy-lift launch vehicle (HLLV) would be needed to launch the various spacecraft elements into low Earth orbit, where they would be assembled. Around 1000 tonnes of payload would have to be lifted into low Earth orbit for each sprint. A total of three sprint missions would establish a permanent outpost on Mars by 2010.

Separate elements

Each sprint mission would involve two main elements flown separately: a cargo vehicle and a piloted craft. The cargo vehicle would be sent to Mars well ahead of the crew, carrying everything they would need on the surface, plus the fuel for their return. Once the cargo vehicle had arrived safely in Mars orbit, the piloted vehicle would depart for Mars, eventually docking with the cargo vehicle to refuel and prepare for landing.

After 10 to 20 days on Mars, the astronauts would return to orbit to dock with the return vehicle. They would head back to Earth and spend time in a rehabilitation facility on the US Space Station.

The sprint missions would require development of new techniques such as aerobraking (see page 81), efficient propulsion systems, storage and transfer of cryogenic (low-temperature) liquid propellants, and robotic systems.

A driving force behind the development of the sprint concept has been the need for less costly options for achieving a Mars landing. To this end it has been suggested that the HLLV could be built out of existing Shuttle hardware. The Orbital Transfer Vehicle (OTV) that will be generally needed as a space tug in Earth orbit could be the basis for the sprint missions' cargo and piloted vehicles.

Earth recovery operations

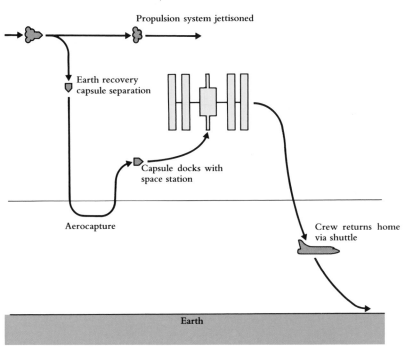

Propulsion system jettisoned

Earth recovery capsule separation

Capsule docks with space station

Aerocapture

Crew returns home via shuttle

Earth

PART TWO
THE HUMAN FACTOR

The first voyagers to Mars will almost certainly be military men, probably aged between 45 and 55. All will have had experience in space already. They will have specialized backgrounds, and at least one will be a doctor. Stability in their personal lives will be all-important: they will probably be married—but with no dependent children.

Why should they be drawn from the military? The simple answer is tradition. Both the Russians and the Americans chose their early space crews from the armed forces. The military are expert at selecting, and then training, people to do the seemingly impossible: to face danger as an everyday experience, and yet to continue functioning normally without loss of efficiency.

So why are women not likely to be included? After all, women can often cope with the demands of getting along with others under exceptionally stressful circumstances. Yet it is significant that the Soviet Union, which has encouraged women to compete with men in almost all other spheres, has been reluctant to send women into space for long flights. Indeed the country with the greatest number of man-hours in space

(far more than the US has achieved) has included only two women, Valentina Tereshkova and Svetlana Savitskaya, among over 100 cosmonauts. And though there may be more female cosmonauts in training, it is unlikely that they will be part of the first Mars crew, because of relative inexperience.

Although the American military will not allow women on combat duty, NASA has been less inhibited than the Russians about sending women into space. Mixed crews have at least been con-

US astronauts training inside a modified Boeing 707, known as the 'vomit comet' because of the steep dives it makes to produce brief periods of weightlessness.

sidered for long trips such as the Mars mission. There could be obvious advantages in a mixed crew—especially teams containing husbands and wives—so long as pregnancy did not result. Even so, there are problems, such as the need for additional facilities for personal hygiene and privacy. Avoidable extra payload weight is always frowned on.

There aren't many reference points to guide selection for a long-duration voyage. On the American side there have been the Apollo astronauts and the crew that accomplished the longest US flight, 84 days in Skylab in 1973–4. On the Russian side cosmonauts have achieved much longer flights—over six months at a time, cooped up inside a fairly small space station in Earth orbit.

Perhaps the most testing mission of all was the abortive Apollo 13 flight. The spacecraft was ruptured by an explosion on the way to the Moon, and the crew was unable to turn back. They endured extreme danger with laudable composure while they flew round the Moon and back to Earth. But they managed to get back safely within a few days, whereas a Mars crew that got into trouble could be years away from home.

Cut off from mankind

In the past explorers have faced similar demands. When Columbus, or Vasco da Gama, Magellan or Henry the Navigator set out on voyages of discovery they had much less idea than a Mars astronaut of how long they would be away. They knew nothing of the problems they might confront, from the elements or perhaps from a rebellious crew; and they had no idea how hostile might be the people or the environment of the lands they were seeking. Arctic and Antarctic explorers have coped with similar odds for very long periods, cut off from any contact with the outside world, with the added stress of subzero temperatures.

So the challenge of the Mars journey, though of epic proportions, is not completely without precedent. But how can space travellers be trained to rise to that challenge? Some experts believe that it will not be possible to train such crews, and that the psychological problems of confinement will be too great.

NASA has not yet faced this problem. It has spent most of its efforts since the Apollo program on short-duration, near-Earth exploitation of space, mostly in connection with the Space Shuttle. NASA hopes eventually to use the Shuttle to build the Space Station in Earth orbit, when it can start gaining the important practical knowledge of how to keep people in space for long flights. But even then it does not envisage flights lasting more than about three or four months for an individual astronaut. In any case, after the delay to the Space Station resulting from the Challenger disaster, NASA has not felt the same need to perfect techniques of training to cope with long flights.

The Soviet Union, on the other hand, has been putting its cosmonauts through intensive training to ready them for long-duration flights, and something is known about the extraordinary lengths to which they go.

Realism in training

General Beregovoi was formerly in charge of Soviet cosmonaut training. He believes in the maximum possible realism and sets training tasks that could kill a cosmonaut. As he puts it:

'*when a man realizes a mistake might cost him his health or even his life, the situation is enormously different.*'

Each trainee cosmonaut has to make at least a hundred parachute jumps, each tougher than the last. This familiarizes the trainee with weightlessness—and also tests his nerve. Towards the end of the series, many of the jumps start with a period of free fall. The free fall has to be long enough for the trainee, using a radio, to spell out a detailed checklist and then to answer questions about ground features—before, that is, being allowed to release his parachute. If he fails even once it could cost him his chance ever to fly in space.

Cosmonauts are given other very tough survival tests. Tests like being carried inside a Soyuz capsule, usually slung beneath a helicopter, and put down in some remote location, often in bitter cold. They are then left, sometimes completely alone, to survive for days without any contact. There is no chance of seeking help: no rescue team is around, and they know it.

Soviet cosmonauts during training for an emergency landing in the sea.

Soviet cosmonaut Valeri Ryumin, who has flown in space five times.

Some trainees have undergone the most amazing physical and psychological hardships and come through. One team, trying to find its way back, fell into a frozen lake but somehow managed to stay alive and still had enough staying power to find its way to safety.

Space crews need the courage and resourcefulness to keep on wrestling with problems long after an average person, given a chance of rescue, would have taken it. When two cosmonauts on Salyut 7 suddenly lost most of their fuel through a fractured pipe, newspaper headlines screamed that they had only two days to live. Yet they gave up the chance of an immediate return to Earth and courageously stayed on board to fix the fault—an almost impossible and highly dangerous task. But they accomplished it and remained in Salyut for several weeks, until their mission had been due to end anyway. It is that kind of resilience, together with highly developed technical skills, that will be required of travellers to Mars.

Medical risks, not technical problems, are likely to prove the biggest obstacles to a manned mission to Mars. Facilities to treat sick crew members will be severely limited, and there is no question of replacing incapacitated personnel. So their health and fitness must as far as possible be guaranteed in advance.

But medical science has never before faced such questions as: 'How do you know that this person will stay fit for the next three years? How do you guarantee that he won't have a heart attack, develop a kidney stone, get appendicitis? . . . '

Crew selectors will have to look into family histories to try to determine the risk of an astronaut contracting an hereditary disorder. A check will be needed to try to ascertain the risk of genetic heart disorder or other dormant threats that could prove fatal or, at best, incapacitate the individual.

But even for the healthiest, weightlessness creates many physical problems. Some are relatively slight, like nausea—space sickness—or a temporary loss of body fluids. Others are much more serious, such as bone deterioration, weakening of the heart and other muscles and a lowering of the immune system's defences. Add to this the hazards of radiation, and the risks of a flight to Mars are formidable. Indeed, one has to think of including a morgue on the spacecraft— the prospect is that serious.

Space sickness

For about half the people who go into space, fitness problems start in a comparatively small way at the start of a flight: the doctors call it space adaptation syndrome. It's motion sickness. It lasts only three or four days and so is no real threat to a Mars mission. The nausea is caused mainly by the sudden upset to the balance mechanism of the inner ear, the otolith organ. Normally gravity controls the position of tiny particles of calcium on sensitive hairs in the ear, and these tell the brain what action is needed to keep balance. But in weightlessness the calcium grains take up random positions that baffle the brain. Drugs such as Scopolamine-Dexedrine can provide relief but are not always successful as a treatment. They can also cause drowsiness at the start of a mission—just when crew alertness is essential.

Disturbing visual stimuli, like the Earth seen overhead, can also be a cause of motion sickness, and astronauts are taught to be ready for such views before looking out of a spacecraft window. 'Always know where your horizon is going to be,' they are warned.

A rush of blood to the head

Other problems come from a redistribution of blood. Normally blood is drawn downward by gravity, and the main arteries, which are equipped with sensitive pressure gauges called barorecep- tors, see that the heart pumps just the right amount to the head. When the body is weightless this delicate mechanism is disturbed. The blood flows freely and around 3½ pints (two litres) moves from the legs towards the head. This makes the face puff out and causes nasal stuffiness, not unlike a head cold. This lasts between one and two weeks, after which the body adjusts and feels normal again.

But things are not normal. For one thing, the heart has less to do than normal and so relaxes. The brain thinks the extra blood in the head means there is too much fluid in the body and so it releases a hormone that tells the kidneys to discharge more urine—as much as 10 per cent more. This can cause dehydration.

It is known that the Russians have made their cosmonauts consume about five pints (three litres) extra fluid per day throughout a flight to make up this loss. On short flights there is no need: then the cosmonaut need only take about two pints (a litre) of saline solution a couple of days before re-entry, to increase the blood vol-

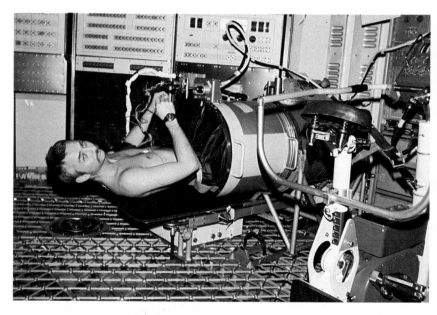

Skylab astronaut Owen Garriott, MD, tests the experimental Lower Body Negative Pressure Device, which was designed to counteract blood pooling in conditions of weightlessness.

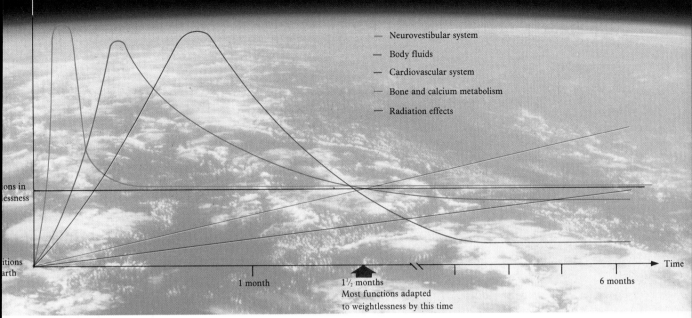

Neurovestibular system
Body fluids
Cardiovascular system
Bone and calcium metabolism
Radiation effects

ons in
essness

itions
arth

1 month   1½ months   6 months   Time
Most functions adapted
to weightlessness by this time

The effects of weightlessness on the human body. The plots show how various physiological functions change on exposure to zero gravity. The neurovestibular system (blue line), the balance mechanism within the inner ear, is affected first: after a few days, it returns to normal. The fluid balance within the body (green line) changes next, and then the cardiovascular system (the heart and blood vessels). After 1½ months all three functions have adjusted to weightlessness. However, the balance of calcium production and absorption in the bones (orange line) and the cumulative effects of radiation (red line) do not adjust. This is the most worrying aspect of long-term space flight: these processes may be irreversible.

ume, and then continue drinking extra fluids for some days after landing.

Muscle wastage

More serious is the discovery that on long flights the heart shrinks in size, by as much as 10 per cent. It is feared that this change could in time become permanent and make the space traveller unable to cope with gravity ever again.

But it is not only heart muscle that weakens: muscles in the legs, especially, tend to get lazy, no longer having to support the body against the pull of gravity.

Once the Russians had developed spacecraft large enough to accommodate pedal exercisers and treadmills, they set their cosmonauts the chore of exercising as much as three hours a day, every day, to overcome muscular decline. This has helped to establish a succession of space endurance records.

NASA, too, will have to face up to the problems of muscle wastage on future long flights. Exercise has played a limited part in its astronauts' daily routine: on Skylab, for example, there were treadmills and pedal exercisers. But NASA is now looking at isotonic and isometric exercises as replacements for such devices.

But exercise has not completely solved the problem. So the Russians have introduced a device to give electrical pulse stimulation to the muscles, and a special garment that demands full muscular exertion from the wearer even to stand upright. The cosmonauts who have to wear it call it a 'penguin suit'.

Countless science-fiction writers prophesied that weightlessness would be one of the rewards of space travel, a liberating experience of release from the bounds of gravity. So it may be, for the first few days; but in the long term it is proving to be the scourge of the space voyager.

Bone deterioration

Possibly the most worrying physiological effect of long-duration spaceflight is the deterioration of the bones. Weight-bearing bones are no longer affected by gravity, and an apparently irreversible loss of bone calcium takes place. It is not fully understood how this happens, but the loss is severe. Most of the calcium lost is discharged in the urine. The

Astronaut Jack Lousma takes a shower on the second Skylab mission.

loss causes concern for several reasons: the bones could become so brittle that they would break easily; and painful kidney stones could develop, formed from some of the calcium lost from the bones.

Most of the calcium is lost from the bones of the legs and arms. Taking drugs to try to increase bone calcium could prove dangerous: the added material might be absorbed indiscriminately by the whole skeleton—producing, say, a thickening of the skull that could damage the brain. But the alternative could be equally disastrous: severe disability for the space traveller for life.

Artificial gravity

It has been suggested that most of the ill effects of spaceflight discussed so far might be overcome by the simple expedient of spinning a spacecraft to create artificial gravity.

There is merit in this idea, but the most widely accepted opinion now is that it could create new problems that outweigh its possible advantages. Apart from new medical worries—people respond in different ways to various speeds of revolution—it is thought that the cost of designing and building a spacecraft would be much higher if it were to be given spin.

And what happens when an astronaut has to move between the outer rim of a rotating spacecraft

Soviet doctors have long taken a keen interest in the medical effects of space travel. They subject cosmonauts to an exhaustive battery of tests before, during and after their flights.

and, say, its hub—perhaps to enter a laboratory where zero-gravity experiments are being carried out? He would almost certainly suffer motion sickness—and this would happen every time he moved from one section of the craft to another. A similar problem would occur if the craft, rather than spinning as a whole, incorporated a small onboard centrifuge.

Bombardment by radiation

Potentially as dangerous as muscle wasting and bone calcium loss is the radiation that space personnel encounter. Radiation is fatal when it comes in doses of hundreds of rems (the name 'rem' stands for 'roentgen equivalent man'). But on the ground the average person gets only about a fifth of a rem per year from the natural background and from artificial sources—we are protected from most of the harmful rays from the Sun and the Galaxy by the Earth's blanket of atmosphere and its magnetic field. Yet space crews orbiting the Earth have been known to receive a

whole rem in the course of a 10-day flight, even though they have been shielded by the Earth's magnetic field.

Beyond the magnetosphere—that is, outside the Earth's magnetic field—the chance of getting a lethal dose of several hundred rems becomes a significant risk. At certain times the Sun releases enormous bursts of radiation—solar flares—which destroy body cells and could kill a man within a matter of hours (see page 74). Shielding would have to be very heavy and many times the thickness of, say, the Space Shuttle's. The best compromise seems to be the provision of a 'storm shelter', a strongly protected area to which the crew could retreat during any period of intense radiation—assuming they received sufficient warning.

In addition to solar radiations, space is filled with cosmic rays, which come from elsewhere in the Galaxy. These, too, are harmful. The seriousness of the effects depends on which cells are struck

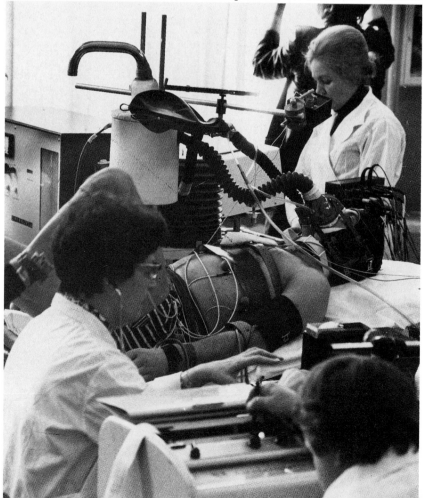

and how severely they are damaged. Shielding from cosmic rays is almost impossible. The extent of this danger is not fully known at present.

Trouble with free radicals

The Soviet Union has greater experience than the West with manned spaceflight: by the beginning of 1987, its cosmonauts had logged a total of 12 man-years in space, compared to only five man-years by US astronauts. Though Soviet space doctors would hardly claim to have solved all the physiological problems of long-term flights, that experience has been invaluable.

An important area of investigation is the role in the body of the energetic molecular fragments called 'free radicals' (see box). According to Dr Anders Hansson, a highly respected biophysicist at the private London-based consultancy Commercial Space Technologies, this area of study is promising for long-term space flight. Soviet research has shown that the normal delicate balance between the natural production and absorption of free radicals is seriously disturbed in space. This change plays a part in the ill effects of spaceflight that we have already noted, from the disturbance of the balancing mechanism in the ear and the loss of bodily fluids to the degeneration of bones, and is aggravated by exposure to extra radiation.

The Russians try to counteract dehydration by making their cosmonauts drink more water, because water counterbalances the activity of free radicals. They also give them extra vitamins and foods that contain 'free-radical scavengers'. They are also investigating certain free-radical enzymes that affect the calcium in the bones.

None of these hazards of space travel will deter men from going to Mars. But they are going to make the difficulties of survival that much greater.

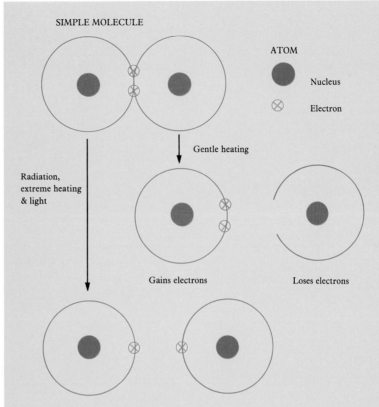

Upsetting the balance

Bombard any matter with enough energy—heat it, for example—and it will start to break up. The delicate electrical balance of the atoms within a molecule will be altered when this occurs. The atoms will acquire either a surplus or a deficiency of electrons, the particles of negative electric charge present in all atoms (an electric current consists of electrons in motion).

Free radicals are highly energetic fragments of molecules, produced when these are broken down by more extreme means, such as very high temperature, very intense light, or radiation. The free radical is a molecular fragment left with a 'rogue' electron, which makes it highly reactive and unstable.

Free radicals are necessary for normal metabolic processes in the body. But if there is an excess of them, the metabolic system cannot cope with their extra energy. An everyday example of free radical damage is butter turning rancid when left in sunlight.

Central to many essential bio- chemical processes is the very strong chemical bonding between hydrogen and other elements. It ensures that, for example, DNA keeps its double-helix shape intact: this allows the genetic code that it carries to be reproduced in the correct sequence. Free radicals, however, are highly oxidizing, which means they have a very strong affinity for hydrogen. They will readily strip hydrogen atoms from nearby molecules, and this can develop into a chain reaction. In living organisms, free radicals have more than enough energy to break hydrogen bonds: as a result they can cause havoc in living cells. Water is a useful antidote to free radical damage because of the hydrogen atoms it contains.

If it is unchecked, the oxidation caused by free radicals can be severe enough to result in death for a living organism. Free radical damage has been implicated as a major factor in the ageing process and has been proved to be linked with cancer. Free radicals may also be a contributory factor in many degenerative diseases such as heart, liver and lung diseases.

PSYCHOLOGICAL HAZARDS

There is such a thing as a fear of space, just as there is a fear of flying. Some astronauts, in their more private conversations, admit as much. Perhaps the most dramatic public evidence came, surprisingly, from the Soviet space programme. During a mission in Salyut 7 in 1985, Vladimir Vasyutin showed worrying signs of strain after being almost two months in space. A fellow cosmonaut, Viktor Savinykh, said he was a 'bundle of nerves', and that there was cause for concern. Within three days, Vasyutin was returned to Earth, where he was diagnosed as mentally unstable and put into hospital.

But this fear of space differs from fear of flying in that it manifests itself during a flight, not before. And that makes it a problem for the planners of a Mars trip. They want to know in advance what the effect on a crew will be of utter isolation in the vast emptiness of space, when the Earth has shrunk to a mere pinpoint of light and Mars is still just a dot in the sky. How will people react to being cut off from direct contact with the Earth for so long a time? How will a group get along together? How will individuals behave?

The United States and the Soviet Union have followed different approaches to the psychological needs of those who go into space. This partly stems from the insistence by US astronauts in the early days that they did not need psychologists, whom they regarded as too theoretical to contribute to their own intensely practical concerns. NASA carried out no systematic psychological debriefings of the Apollo or Skylab crews—but now they are having to think again.

But psychology has not achieved any remarkable breakthrough in the control of human behaviour, nor can we expect it to. People just aren't that malleable. *Homo sapiens* has been around a long time but still he fights, gets irritable or frightened, suffers anger or the torment of loneliness.

Cosmonaut Vladimir Vasyutin, who became mentally unstable after a few weeks in space.

So far NASA's clues to how people might behave during long periods of isolation have had to come mainly from studies of groups in the Antarctic, or crews confined in nuclear submarines that stay submerged for months at a time. Psychologists say there have been 'remarkably consistent' findings: sleep disturbance (three quarters suffer from this), bore-dom, restlessness, anxiety, anger, depression, headaches, irritability, a loss of the sense of time and space, and a general lowering of concentration.

The Soviets, with their greater experience of long-duration flights, have been much more systematic in their psychological studies. Their special Group for Psychological Support carries out research, and its psychologists help select and train the crews, make observations of their mental state throughout each flight, and carry out a full debriefing afterwards.

Communication problems

But earthbound research cannot embrace what are bound to be huge psychological differences among people who will know they are completely cut off from Earth itself—where even radio signals, travelling at the speed of light, could take over 20 minutes to cross the vast distance that separates them from home.

All long-duration flights so far have been in Earth orbit, where crews have been able to keep in regular, instantaneous contact with loved ones below. But on a Mars flight it will be different. Two-way conversations will not be easy to set up, or to conduct when they do take place, because of the long delay between the two sides of the dialogue. The crew will have to rely largely on occasional prerecorded messages from family and friends.

Care will have to be taken with the serious problem of bad news from home. During a 96-day flight in Salyut 6, the father of Georgi Grechko, one of the cosmonauts, died. The information was relayed to his partner, Yuri Romanenko, who decided not to tell Grechko until the flight was over. Grechko agreed afterwards that this had been the right thing to do.

But what would happen on a Mars trip if someone suspected that bad news was being kept from him as a matter of policy? The very uncertainty could cause torment, and paranoid feelings prompted by just such a cause may have contributed to Vasyutin's breakdown. So it might be better to promise the crew beforehand that they will always be told the truth. Psychologists admit, however, that they have yet to decide on this.

Who's in charge?

The communications delay also means that the Mars crew will have to control most, if not all, of the mission themselves and not rely as much on people on the ground as in the past. What then, should be the degree of authority vested in any one crew member? Should the commander have an autocratic role or should all members of the crew have a say in running the ship?

Russian cosmonauts have almost no hierarchical roles, whereas the Americans ensure that their astronauts have them. But the US Aerospace Medical Association says that 'a Mars crew would need to behave as a family with the interests of the group prevailing over individual interests . . . a family strongly supportive of one another.'

A compromise will probably be reached: the commander taking autocratic powers when it comes to operating the ship itself, the crew voting democratically on how to solve social problems that arise from living together in a confined space for so long. There is after all, nowhere to go to 'get away from it all', though this could be mitigated with the provision of 'private' cabins.

A problem could arise in later flights when crews come from both military and scientific backgrounds, with friction over priorities if the mission has goals of both types. Hitherto, military space operations have been separated from civilian and scientific ones, but the cost in money and time of going to Mars might suggest combined missions as a matter of expediency.

The strain of only a few days in space is reflected in astronaut Donn Eisele's tired face, photographed on Apollo 7 in October 1968.

The alchemy of space

It is frequently suggested that people who have been in space—especially those who went to the Moon—have been profoundly changed. It is often said, for example, that one Apollo astronaut, Jim Irwin, 'turned to religion because he thought he saw God on the Moon'.

These stories are largely myths. Take the case of Jim Irwin. He was a highly religious man before he became an astronaut and it was to be expected that he should feel the presence of God while standing on the Moon and looking towards Earth. Similar thoughts have crossed the minds of all astronauts with a normal religious upbringing when they have viewed the Earth from space.

Yet it is true that toward the end of the third Skylab mission, which clocked 84 days in space, the crew went on strike. Even Ground Control (according to Henry Cooper in his book *House in Space*) began to wonder: 'Was there something in the strange alchemy of space that had changed the characters of the astronauts? They were unaccountably irritable . . . they complained . . . they bitched . . . they grumbled.'

At the end of the sixth week in space the whole crew took a day off as a kind of act of rebellion, a 'declaration of independence'. Such behaviour, it was realized, could jeopardize a mission. It led astronaut Ed Mitchell to warn that 'we may see a bunch of mental dropouts when people start flying into space by the hundreds if we don't start to learn how to prepare them for the experience.' And it caused people to give new respect to the NASA psychologists who were trying to solve some of these problems.

The workload dilemma

Boredom, and all that can stem from it, is one of the main concerns of the psychologists. All the crew must be given absorbing tasks to keep them occupied—but not to such a degree that they come to resent it.

Cosmonaut Vladimir Kovalyonok, who once spent 139 days

in Salyut, said: 'Rest and work in space have different meanings. In space you want to load yourself with work to make the time go faster. Otherwise you feel the loneliness.'

But the other side of the coin is fatigue. Jack Lousma said after his flight in Slylab 3: 'The things that suffer when you want to get something done or you're running behind are meals, exercise and getting to bed on time. Yet these are the three highest priorities that you need to do on time and regularly.'

And Soviet space officials have said that quarrels between cosmonauts, which often start after about 30 days in space, are usually caused by fatigue.

So irritability is a major problem. There are ways to lessen the damage it causes: by the careful selection of food to suit each taste, by the design of colour schemes for the spacecraft's walls and equipment, by the provision of music of each individual's choice. Soviet cosmonauts have even been given recordings of the sounds of home—birdsongs, waves on a sea-shore, the sound of falling rain, that sort of thing.

It will be necessary on future long flights to provide substitutes for the normal clues to the passage of time—of which the most important are day and night. It is recommended that lights be automatically dimmed as 'evening' approaches, finally being turned down to near-darkness to simulate night. This will help maintain normal daily rhythms. Other advice from NASA is that during the course of the flight several intermediate goals should be celebrated on board, to engender the feeling that progress is being made. 'In isolation', they point out, 'days tend to blend into one another.'

The sex urge
Finally, perhaps the most important psychological problem of all—how to deal with the normal sex urge of adults. Whatever solution

Above: Mir provides cramped living quarters for up to six cosmonauts, but in weightlessness extra space is available near the ceiling.

Below: the first six-man crew in history (Spacelab 1, 1983) strikingly illustrate the fact that in space there is no 'up' or 'down'.

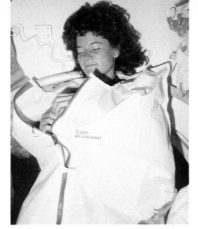

Right: Sally Ride demonstrates the comforts of sleeping clipped to the Shuttle's cabin wall. An inflatable sleeping bag simulating the effect of gravity has also been tried out.

Left: privacy is essential on long space flights. The Soviet Mir space station has two individual cabins, each one equipped with a porthole and a 'vertical' sleeping bag.

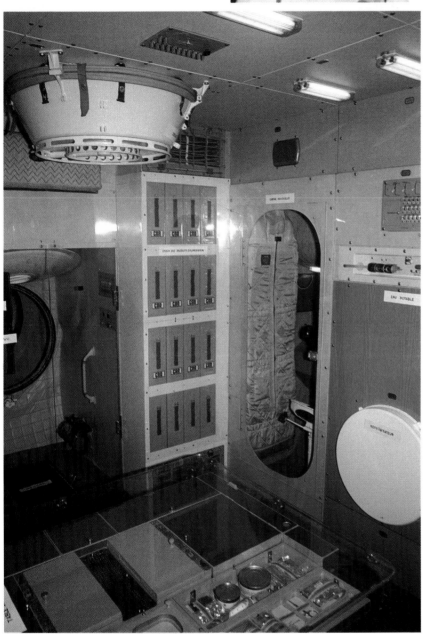

Above: Mir's living quarters have a dining table, a shower unit clipped to the ceiling, a white refrigerator at the far right, a food cabinet at the rear and individual cabins.

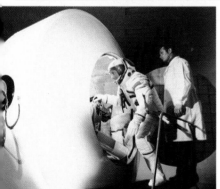

Left: space travellers are prepared for extreme accelerations by spells in a centrifuge, in which a cabin mounted on an arm rotates at high speed. Shuttle astronauts feel three times their normal weight during launch.

is proposed seems to contain its own weaknesses. When NASA officials were first questioned on this, they would go only so far as to say 'close coupling of crew members is to be avoided' (without defining these terms!)

Were they envisaging an all-male crew and, if so, were they trying to say that homosexuality would be ruled out? Did they foresee the possibility of a mixed crew and, if they did, would they insist on couples among the crew being married? If so, what of the conflicts that would arise from the different treatment of unmarried crew members?

If, on the other hand, 'free love' were accepted in a mixed crew, what of the dangers of jealousy, or depression resulting from rejection? The mission could be jeopardized if even one member of the crew were so affected.

With NASA unwilling to comment (the subject is virtually taboo), a group of leading space psychologists offered this statement:

We suggest that a sexual relationship between crew members or, more importantly, a special emotional relationship, would be potentially harmful to the stability of the crew as a whole and their ability to provide support to all members throughout the length of the mission. Therefore we recommend the avoidance of such relationships, at least until more experience is gained with long-duration spaceflight. If the crew fully understands the potential for negative impact—jealousy, special treatment, circumvention of the chain of command, recriminations and regret if a relationship fails, and so forth—we believe they will adopt this avoidance policy as a group norm, which will be enforced by expectations and positive pressures during the training and the mission.

Stripped of the jargon, this could be paraphrased as: the very best of friends can become the deadliest of enemies. Privately the psychologists admit that this is a major problem with no easy answer.

PART THREE
JOURNEY AND ARRIVAL

TRAJECTORY TO MARS

It is impossible to go directly to Mars. The energy that would be required would be far too great, even for today's advanced rocket engines.

In space there is no such thing as a straight-line path. The planets and their satellites all travel in elliptical orbits. A spacecraft, coasting from one planet to another under the predominant influence of the Sun's gravity, traces out a long elliptical trajectory.

This would be complicated enough to calculate in advance, but there are also disturbing influences. Isaac Newton was the first to realize, in the 17th century, that every object exerts some measure of gravitational pull on every other. Over various stretches of its journey, a spacecraft travelling from the Earth to Mars will be affected by the gravitational influences of both those planets and, to some extent, by those of all the others.

Calculating the resulting trajectory involves extremely abstruse mathematics. But back in the 1920s, when astronautics was still a disreputable science in the eyes of many, a number of pioneers had the tenacity and patience to solve the complicated equations involved.

The transfer orbit

In 1925 a German engineer named Walter Hohmann hit on the key to future interplanetary travel. As we have noted, a spacecraft leaving Earth would inevitably describe an elliptical orbit around the Sun—just as if it were another planet. Adjust that ellipse at the start and you have the easiest way to send a craft inwards, to Venus or Mercury, or outwards, towards Mars and the outer planets.

Hohmann showed how the best trajectory for a spacecraft could be chosen—the one most economical in terms of both energy and flight time. In the case of the Earth–Mars journey, the trajectory would lie outside Earth's orbit, though just touching it at the launch point, and inside the orbit of Mars, just touching it at the rendezvous point. The spacecraft would arrive at its destination at precisely the opposite point in its orbit from its launch position (see figure).

The space age has proved Hohmann right. Hohmann transfer orbits, as they are now generally known, have been used for interplanetary probes for decades.

The master stroke of Hohmann's pioneering work was his realization that a spacecraft could change its direction without expending extra energy: instead, the gravitational pull of the Sun could provide that.

Launch windows

It is impossible to launch a spacecraft at will. Departure time has to be calculated accurately. Mission planners need to know the precise size of the payload so that they can boost the rocket to the exact speed that will bring it to the desired point in space, millions of miles away and months later—at the same moment that Mars gets there, too.

The time of launch must lie

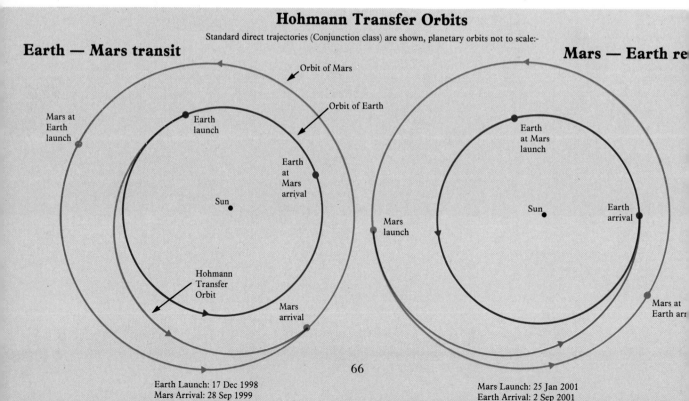

Hohmann Transfer Orbits

Standard direct trajectories (Conjunction class) are shown, planetary orbits not to scale:-

Earth — Mars transit

Orbit of Mars

Orbit of Earth

Mars at Earth launch

Earth launch

Earth at Mars arrival

Sun

Hohmann Transfer Orbit

Mars arrival

Earth Launch: 17 Dec 1998
Mars Arrival: 28 Sep 1999

Mars — Earth re

Earth at Mars launch

Sun

Mars launch

Earth arrival

Mars at Earth arr

Mars Launch: 25 Jan 2001
Earth Arrival: 2 Sep 2001

The last sunrise over the Earth seen by the Mars explorers before accelerating out of orbit.

Mars mission classes

In the 1920s Dr Walter Hohmann discovered the most energy-efficient route between the planets (see main text). But with greater expenditure of energy other paths can be followed. Trajectories to Mars can be divided into two types of orbiter/lander mission: 'opposition' and 'conjunction' classes.

Paths of the opposition class permit a stay at Mars of only 60–80 days before the planets move out of the right alignment for the return journey. A Venus flyby, with its attendant disadvantages (see main text), is needed on one leg of the journey. The mission lasts two years overall. The schedule for such a flight, arriving at Mars in 1999, could be:

Earth departure	26 Jan 1998
Venus swingby	9 Jul 1998
Mars arrival	16 Jan 1999
Mars departure	17 Mar 1999
Earth arrival	18 Nov 1999

The conjunction class of missions is more energy-efficient because the spacecraft waits longer on Mars until a more favourable Earth return window comes around. No Venus swingby is needed but a serious drawback is that the wait on Mars could be 300–550 days. The total mission length would be three years. For a turn-of-the-century mission, the schedule could be:

Earth departure	17 Dec 1998
Mars arrival	28 Sep 1999
Mars departure	25 Jan 2001
Earth arrival	2 Sep 2001

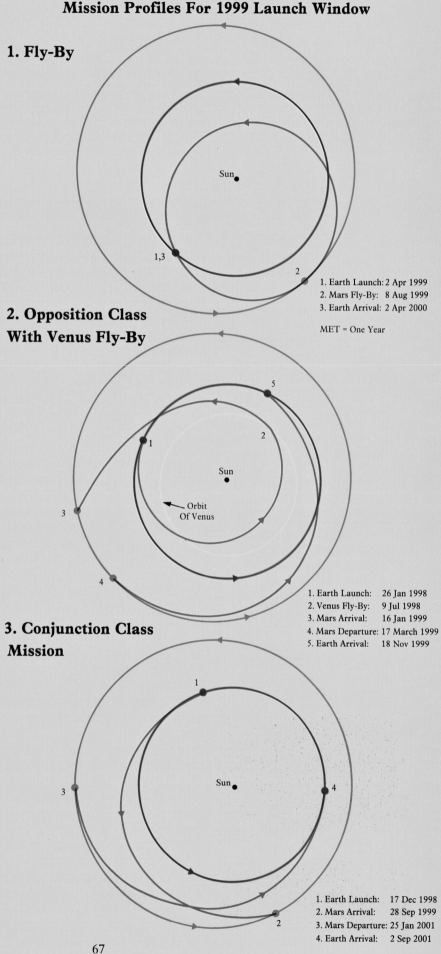

Mission Profiles For 1999 Launch Window

1. Fly-By

Sun

1,3

2

1. Earth Launch: 2 Apr 1999
2. Mars Fly-By: 8 Aug 1999
3. Earth Arrival: 2 Apr 2000

MET = One Year

2. Opposition Class With Venus Fly-By

5

1

2

Sun

Orbit Of Venus

3

4

1. Earth Launch: 26 Jan 1998
2. Venus Fly-By: 9 Jul 1998
3. Mars Arrival: 16 Jan 1999
4. Mars Departure: 17 March 1999
5. Earth Arrival: 18 Nov 1999

3. Conjunction Class Mission

1

3

Sun

4

2

1. Earth Launch: 17 Dec 1998
2. Mars Arrival: 28 Sep 1999
3. Mars Departure: 25 Jan 2001
4. Earth Arrival: 2 Sep 2001

somewhere within the Mars launch window. Imagine that the sky above you, as you stand on Earth, isn't transparent but opaque, blotting out any chance of reaching a chosen destination. Every so often a gap or 'window' passes over the Earth through which a rocket can be launched with the certainty that a target will be at the other end of the flight to greet it. Delay too long and the 'window' closes. With Mars the opportunity occurs roughly every 25 months.

Escape velocity

Right from the launch natural forces are enlisted to ease the energy requirements of the mission. The Earth's rotation can be used to 'sling' a rocket into space. Places on the equator are moving fastest—at well over 1000 miles per hour (1600 kilometres per hour). That's why some launch sites are close to the equator—to get maximum free thrust from the Earth.

Even so, a rocket has to build up a speed of about five miles per second (eight kilometres per second) just to get a payload into Earth orbit. This is a speed of 17,500 miles per hour (28,000 kilometres per hour). But if the payload is to be propelled away from Earth, a final boost is needed to push the speed up to seven miles per second (11 kilometres per second). This speed of 25,000 miles per hour (40,000 kilometres per hour) is known as 'escape

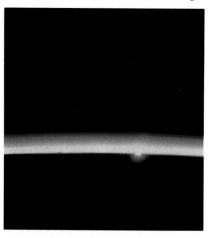

velocity'. When it has been attained, the rocket can shut down: the Earth will gradually slow the spacecraft, trying to drag it back again, but it will never quite succeed in bringing it to a stop. If it is Mars that the spacecraft is travelling towards, it will eventually reach the point where the pull of Mars becomes dominant; then it's 'downhill' all the way.

Gravity assist

Another powerful gravitational trick used in interplanetary travel is known as 'gravity assist' or 'planetary swingby'. By using the gravitational influence of a planet it is possible to visit others without the need of additional rocket thrust (see box). This technique has allowed the Voyager spacecraft to visit nearly all the outer planets and Mariner 10 to visit both Mercury and Venus.

Gravity assist could be used for manned missions to Mars. Strange though it may seem, a spacecraft could head inwards towards the Sun and use the gravitational field of Venus to provide the energy needed to fly to Mars. NASA's Galileo mission to Jupiter will use the method, flying a complicated trajectory known as VEEGA, for 'Venus Earth Earth Gravity Assist'. The probe will be launched from the Space Shuttle in late 1989, will fly past Venus and then will pass the Earth again before heading off towards Jupiter.

But there would be disadvantages for the crew if a manned Mars mission were to do something similar. The spacecraft would have to cope with a greater temperature range: near Venus the Sun is about four times hotter than it is at Mars. Payload space would be sacrificed for a cooling system that would have no function at Mars. There would also be problems with the increased level of radiation streaming from the Sun. This would increase the hazards to the crew, and greater shielding would be needed.

Mars launch windows

The combined orbital motions of the Earth and Mars are such that they come into the appropriate positions for a spacecraft launch every 25 months or so. The quality of these windows varies because the distance of Mars from the Sun varies by about 27 million miles (43 million kilometres). It is more economical to fly in years when the Earth–Mars distance is least, intercepting Mars when it is at its perihelion—the closest point in its orbit to the Sun.

These favourable windows recur every 15–17 years, and almost exactly every 284 years. 'Vintage' years for Mars missions include 1956, 1971, 1988 (chosen for the launch of the Soviet Phobos probes) and 2001. Since Mars is close to the Earth during these years, it appears as a bright object in our night skies.

There are two sorts of Hohmann transfer orbits that can be followed. Type I paths have a short flight time, but require extra fuel for braking when the spacecraft enters orbit around Mars. A table of Type I launch windows up to 2005 is given here.

Type II trajectories take the spacecraft on slower journeys to the planets, and more payload can be accommodated because less fuel is required. The Soviets have indicated that their unmanned rover and sample return flights will use Type II orbits. The launch windows occur four to six weeks earlier than the dates given here, and the Mars arrival times fall six to eight weeks later than the corresponding dates in the table.

Launch	Arrival
Jul 1988	Jan/Feb 1989
Sep 1990	Apr/May 1991
Oct/Nov 1992	Aug 1993
Nov 1994	Aug 1995
Dec 1996	Aug 1997
Feb 1999	Sep 1999
Apr 2001	Oct 2001
Jul 2003	Feb 2004
Sep 2005	May 2006

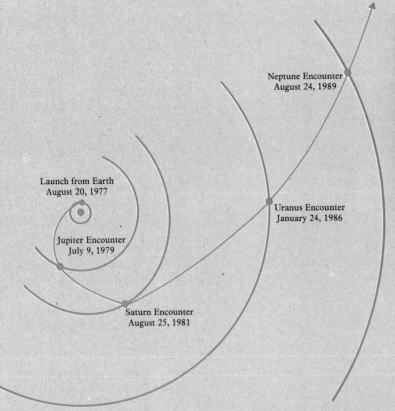

Launch from Earth
August 20, 1977

Jupiter Encounter
July 9, 1979

Saturn Encounter
August 25, 1981

Uranus Encounter
January 24, 1986

Neptune Encounter
August 24, 1989

NASA's Voyager used the gravity of Jupiter and Saturn to reach the outermost planets.

Left: Voyager 2's path during the Saturn encounter in August 1981 was carefully planned to swing the probe around sharply for Uranus flyby 4½ years later. Following the Neptune encounter in August 1989, Voyager 2 will leave the solar system altogether.

Swing your partner

The technique of gravity assist has been likened to a game of inter-planetary billiards. A spacecraft dives in close to a planet so that its path is altered by the gravitational field and it speeds off around the Sun, getting a free ride to other targets. The probe effectively enjoys the benefits of an additional rocket stage at no extra cost.

Since the solar system's energy books have to be balanced, what the probe gains the planet loses. For example, as the Voyager probes were being accelerated by Jupiter, the planet's rotation was being slowed: Jupiter's 'day' was lengthened—by a billionth of a second.

Multi-planet orbits were first considered briefly by theoreticians in the 1920s, but real progress came in the 1950s when increasingly powerful computers could handle the huge calculations required. In the early 1960s NASA's Jet Propulsion Laboratory made the first detailed analyses, and it was soon discovered that probes could reach Mercury via

Venus following launches in 1970 and 1973. A direct journey was possible using the powerful Titan 3C booster, but a gravity assist from Venus would allow the use of the much cheaper Atlas–Centaur, which could not reach Mercury unaided. So Mariner 10 flew by Venus in February 1974, and the planet's gravitation caused it to lose speed and swing in towards Mercury's orbit, which lies closer to the Sun.

One difficulty of the method is the accuracy of guidance required: Mariner's position at Venus had to fall within a 250-mile (400-kilometre) slot, or else it would have missed Mercury altogether. In the event, guidance was so accurate that Mariner 10 arrived within 27 miles (43 kilometres) of the allotted position.

The most dramatic use of gravitational assistance has been by the Voyager spacecraft on their odyssey through the outer solar system. Mathematicians discovered in the 1960s that every 176 years the positions of the outer planets were such that spacecraft

could pass them in sequence without impossibly large fuel supplies. The Voyagers were launched in 1977 on a path that was to take them on an exquisitely timed dance around the outer planets. Voyager 2, for example, reached Jupiter in July 1979, Saturn in August 1981 and Uranus in January 1986. It is now heading towards Neptune, which it will reach in August 1989.

So gravitational assistance is a standard feature of deep space missions. NASA's Galileo probe will use gravity assist to visit the four largest moons of Jupiter in succession. In the early 1990s the European Space Agency (ESA) craft Ulysses will use Jupiter's gravity to swing its path back in towards the Sun while soaring high over the orbits of the planets. Gravity assist can also give space-craft a new lease of life: for example, the paths of the three highly successful Halley's comet probes launched by ESA and Japan may be altered by the Earth's gravitational field in 1990 to send them towards new targets.

THE JOURNEY

Once their main engines have shut down, the Mars voyagers will be travelling faster than anyone in history. Only NASA's quartet of Pioneer and Voyager probes to the outer planets have travelled faster. The travellers will cross the Moon's orbit after less than 12 hours, and will then be more distant from their home planet than any human beings have ever been.

At 900,000 miles (1.4 million kilometres), Earth's clouds and blue-green oceans will still be clearly visible, but its disc will have dwindled to the apparent size of the Moon as we see it from the ground. Around 25 million miles (40 million kilometres), an astronaut will be able to stretch out an arm and cover both Earth and Moon with a thumb.

The Earth and Moon will at first lie off to one side of the Sun, and the crew will see them only as crescents, diminishing in size. Gradually the Earth–Moon system—moving faster than the ship around the Sun—will pass in front of the Sun and then out on the other side. Off-duty crew will probably spend hours gazing at the pair through telescopes, just as the favourite relaxation of the three Skylab crews in 1973–4 was to watch the endless variety of the Earth's surface gliding silently past below them.

The first few days outward-bound will be the crew members' busiest until Mars arrival. Together with ground control, they will be measuring their orbit accurately. If the boost stage has gone wrong, they could fire the engines again to return home. Later their options will be more

View from the Spacecraft

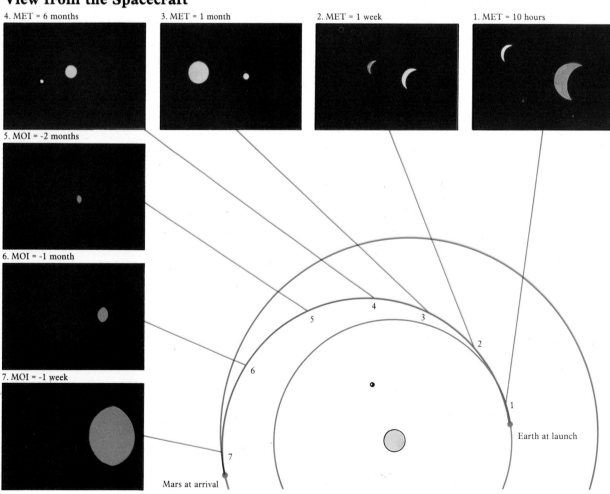

4. MET = 6 months
3. MET = 1 month
2. MET = 1 week
1. MET = 10 hours

5. MOI = -2 months

6. MOI = -1 month

7. MOI = -1 week

5

4

3

2

6

1

Earth at launch

7

Mars at arrival

A few hours after the Mars mission leaves Earth orbit (1), both the Earth and Moon appear as twin crescents, gradually diminishing in size and brightness as the days pass (2). After a Mission Elapsed Time (MET) of a few weeks, the Earth looks like a bright star almost lost in the Sun's glare (3). Eventually it passes in front of the Sun and reappears on the other side (4). A few months before Mars Orbit Insertion (MOI) Mars looks like a red star continually growing brighter (5). As the craft approaches, Mars grows to a disc (6) and eventually fills the sky (7).

limited. It might be necessary to return by flying via Mars, just as the ailing Apollo 13 had to fly round the Moon in 1970 in order to get back to Earth.

But if all has gone well, the crew will find that their orbit needs only slight adjustment. This first midcourse correction will be performed one to four weeks after departure, because it is more economical of propellants than waiting until the ship is nearer Mars.

Finding the way
Calculating the ship's position and orbit is literally a matter of life and death. Without accurate trajectory information the ship could miss Mars—or even hit it. Because it is so critical, several methods will be used. It will even be possible for the crew to do it themselves should contact with Earth be lost.

Most of the Mars and Venus missions in the 1960s were guided by radiometric methods—that is, measuring position by radio techniques. These called on sensitive radio telescopes such as those of NASA's Deep Space Network (DSN) in California, Madrid and Canberra, and the Soviets' comparable instruments in Yevpatoria in the Crimea, Usserisk in the Far East and Medvezhi Ozera in the Moscow region. These telescopes measure the Doppler shift of the radio waves—that is, the change in wavelength due to the motion of the craft transmitting them. (This is analogous to the change in pitch of a train whistle as the train rushes past the listener. The faster the train goes, the greater the change in pitch; the faster the spacecraft goes, the greater the shift in the wavelength of its radio waves.) The DSN is so accurate that NASA is confident of measuring the speed of its 1989 Magellan Venus orbiter with an accuracy of 2.2 miles per hour (3.5 kilometres per hour).

The Doppler shift technique measures speed towards or away from the receiver (but not the

Contact with the Mars mission will be maintained by large Earth receivers, like these Soviet instruments in the Crimea.

'sideways' component), and distance travelled can be calculated from this. Another technique can measure distance more directly: it is calculated from the time a radio pulse takes to travel from the ship to Earth.

But even in combination these techniques are not accurate enough for a manned Mars mission. When the Soviets used them for the VeGa spacecraft in 1985–6, the position accuracy at Halley's comet was only 250 miles (400 kilometres)—far too coarse for Mars. Another method measures very accurately the changing angles between the spacecraft and celestial radio sources such as quasars. This improved the accuracy in the VeGa mission to 25 miles (40 kilometres).

But these techniques are all Earth-based: they become less accurate at greater distances, and what happens if the ship-Earth radio link is broken?

Navigation by eye
The answer is to use optical navigation aboard the spacecraft itself. The technique was first demonstrated by NASA's Mariner 7 Mars

probe in 1969, but not fully exploited until the Voyagers ventured into the outer solar system. Basically it involves measuring the angles between a number of stars—at least three, but accuracy is improved if more are used—and, say, a planet. This is done at a known time, to fix the spacecraft's absolute position in space. It is the same approach that sailors have used for centuries, using a sextant to measure the positions of the Sun or stars above the horizon at known times. Voyager carries no sextant but it can photograph stars and selected planets or satellites together.

Combining these very different optical and radiometric methods allowed engineers to steer Voyager 2 past Uranus with 18.5-mile (30-kilometre) accuracy in January 1986—from a distance of 1790 million miles (2800 million kilometres).

If Earth contact is lost early on, navigational responsibilities will fall squarely on the crew and their computers. The computers will contain the exact celestial coordinates of widely separated bright stars, for comparison with the discs of the receding Earth and Moon. Apollo 11's computer, for example, offered a selection from 37 stars, such as Rigel, Canopus and Capella.

As it makes more and more measurements, the computer will be able to generate an accurate picture of the orbit and the timing of the first midcourse correction. Measuring stellar angles from Mars as it draws nearer will yield increasingly precise orbital data. These will eventually be accurate enough to program the computer for the second and final midcourse correction, one to two weeks before Mars arrival, to 'tweak' the exact position of arrival and braking burn.

Should this navigation computer break down (and there should be at least one backup) the astronauts could measure the angles manually by sextant and do the mathematics,

Interplanetary debris

Space is far from empty: it is full of interplanetary flotsam and jetsam, ranging from dust-sized micro-meteoroid particles to larger chunks of rock. If the smaller particles fall into the Earth's atmosphere, they are burned up, becoming briefly visible as meteors or 'shooting stars'. Sometimes the larger debris survives the high temperatures in its passage through the air and falls to the ground as one or more meteorites.

The Earth is well protected by its blanket of atmosphere. Not so a spacecraft: a collision with an object half an inch (one centimetre) in diameter moving at a speed of 12 miles per second (20 kilometres per second) produces the same energy as an exploding hand grenade. Mission planners have to know the chances of such an impact.

Spacecraft in Earth orbit are in increasing danger of colliding with the remnants of satellites and upper-stage rocket boosters. One estimate suggests that there are now more than 50,000 pieces larger than half an inch (one centimetre) across.

This potential minefield has

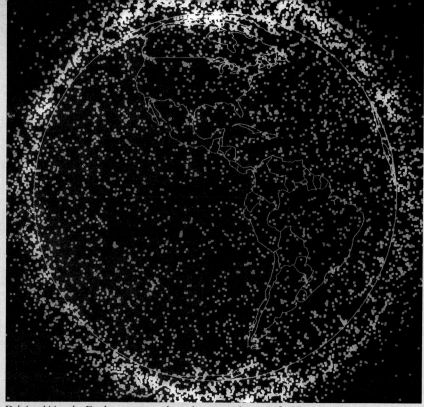

Debris orbiting the Earth now poses a hazard to manned spacecraft. This is NASA's assessment of the 'junk cloud' in 1987.

already taken its toll, though fortunately without catastrophe. A window on the Shuttle *Challenger* had to be replaced after its second mission when a tiny flake of paint from another craft caused a pit a quarter of an inch (half a centimetre) deep. The following month cosmonauts aboard Salyut 7 heard a micrometeoroid slap into a porthole. Parts of the four-year-old Solar Max satellite, brought back

to Earth by Shuttle astronauts in April 1984, showed numerous microscopic pits from strikes by both natural and manmade objects.

The Earth's gravitational field tends to attract debris: beyond Earth orbit there is much less celestial rubbish. The asteroid belt between the orbits of Mars and Jupiter was a matter of concern when missions to the outer planets were planned. In fact Pioneer and Voyager spacecraft passed through it without adverse effects. Pioneer 10 recorded only 55 micrometeoroid impacts between Earth and Jupiter.

More worrying for a manned mission to Mars are the small asteroids that fly free of the main belt and approach the Earth's orbit. Only about 100 are known, but there may be 10 times as many with diameters up to half a mile (one kilometre) across. A systematic search by the Mount Palomar observatory in California is under way: when a manned mission to Mars is planned, a computer will check all the relevant asteroid orbits to ensure that there will be no close encounters.

Dr David Hill and Dr John Zarnecki of the Unit for Space

This Long Duration Exposure Facility, released by the Shuttle in 1984 and still operating, includes experiments to register micrometeoroid impacts.

Sciences at the University of Kent have studied the deep space dust problem for this book. They considered a Mir-type vehicle, presenting a cross-sectional area of 540 square feet (50 square metres) at right angles to the direction of travel. Taking micrometeoroids of all sizes down to a billion-billionth of a gram, they found that during a journey of 259 days a billion strikes could be expected. This sounds horrendous, but the overwhelming majority of these particles are at the small end of the scale: only one strike of a particle weighing as much as a ten-thousandth of a gram is expected. This would have to hit a sensitive area to do any real damage.

The damage from such a hit would be reduced by modern spacecraft design, in which vital equipment is dispersed. The value of this was highlighted by the explosion on board Apollo 13, in which all the main oxygen supplies were lost: later craft were flown with a separate tank well away from the others.

If collision with a micrometeoroid caused air leakage, internal pressure sensors would trigger alarms. A module affected by a large strike would have to be evacuated and sealed off. If crew members wearing pressure suits could not gain access to the hole because fixed equipment was in the way, an EVA (extravehicular activity), or spacewalk, would be needed to patch it from the outside. Pressure suits and EVA suits must not all be stored together: they would be inaccessible if the module they were in should be severely damaged.

Hill and Zarnecki estimate that there is only a 1 in 10,000 chance of being struck by a particle weighing a thirtieth of an ounce (1 gram). Put another way, only one such impact would be expected in 7000 years. But larger spacecraft and large arrays of solar panels will run a proportionately greater risk.

An artist's impression of a Marsbound craft, several days out from Earth. Both Earth and Moon appear as thin crescents.

possibly with the help of a specialized calculator. Buzz Aldrin did something similar in 1966 when the radar on his Gemini 12 spacecraft failed; he fell back on sextant sightings and a set of specially prepared charts to rendezvous with a target—a significant achievement in those days.

Of course, the engines have to be pointing in the right direction when they are fired for these course corrections and for Mars braking. Normally this information is provided by an attitude reference system using gyroscopes that are recalibrated from time to time by star sightings. During the cruise, when exactness of orientation is not so important, sensors lock onto the Sun and a star such as Canopus (traditionally favoured for US unmanned space probes) to enable the attitude control system to hold the ship steady. In the unlikely event of equipment failure, it would be a simple job for the astronauts to position the ship for engine firing by aligning a set of window markings with selected stars. The tiny American Mercury spacecraft had something similar for retrofire from orbit: the astronaut matched a window pattern with Earth's horizon to guarantee he was in the right position.

Cruising to Mars

The time spent crossing the Earth – Mars gulf will be a relatively relaxed one for the crew, once the excitement of course corrections is over. Both they and the ground controllers will keep an eye on the state of the life support systems, the computers and the ship's health in general. Routine maintenance will be interspersed with replacement of some items—spares will be carried for as much critical equipment as possible. If there are any leaks from propellant or engine systems, the crew will be obliged to carry out a series of space walks to repair them, as happened on Salyut 7 in 1984.

There will be time to practise emergency drills for the evacuation of a compartment depressurized by a micrometeoroid strike (see box). This is a routine carried out occasionally on board Salyut and Mir space stations in Earth orbit. The astronauts will relax with books, tapes, chess and family linkups. They will fight off the effects of prolonged weightlessness by exercising for at least two hours per day with a treadmill, pedal exerciser, and chest expander, and carry out regular medical checkups (see page 57).

The long duration of the flight will force the crew to refresh their training, perhaps with the help of instructional videos. Procedures for the critical Mars braking burn will be practised as arrival approaches. By Marsfall, the crew should be thoroughly acquainted with the ship's eccentricities and be fully prepared for the complex work ahead.

Perils of solar flares

The 'solar wind' is the name given to the streams of electrically charged subatomic particles that flood outwards from the Sun, averaging 125–250 miles per second (200–400 kilometres per second). Occasionally there are distinct explosive outbursts of such particles. These are solar flares, and they are a serious threat to deep-space travellers.

Since the middle of the 19th century it has been known that the Sun follows a regular cycle of activity. This was detected by counting sunspots, which were found to vary in number over an 11-year period. When there are few sunspots solar activity generally is low, and weeks can pass without any flares being seen. As sunspots proliferate, solar activity increases, and flares can appear every hour or two. The last solar maximum occurred in 1980/1, and the next will be in 1992/3.

Solar flares appear in close proximity to sunspots. Their origin is still uncertain but they seem to be caused by sudden increases in magnetic field activity in the Sun's outermost layers. When a flare occurs, the energy of 10 million hydrogen-bomb explosions is given out in a few minutes. This produces a shock wave in the solar wind, briefly pumping its speed up to many times the norm. The Sun's outer layers, normally at a temperature of 11,000°F (6000°C), are heated to several million degrees and electromagnetic radiation is generated right across the spectrum. At visible wavelengths small patches on the Sun's surface appear to brighten.

The most energetic flare particles arrive from two to 10 hours after the outburst. They slam into the Earth's upper atmosphere, causing spectacular aurorae and magnetic storms. Shortwave radio communications can be disrupted, and fluctuations in the magnetic field can induce corrosive electrical currents in long pipelines.

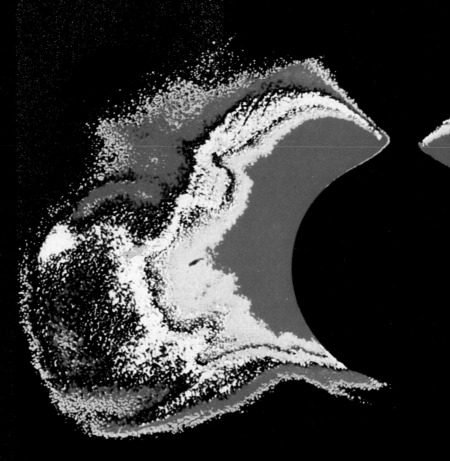

By their very nature solar flares are unpredictable. When Mars is on the opposite side of the Sun from the Earth, flares that would affect Mars cannot be seen by terrestrial observers. So the Mars manned craft will have to contain a battery of solar monitoring equipment. This will include X-ray, radio and optical telescopes, as well as instruments to monitor the Sun's magnetic field. Onboard computer analysis of the data must be performed because of the communications delay between Earth and Mars.

A reliable prediction of a solar flare can be made only 20–30 minutes before increased radiation levels indicate that one has begun. Thereafter only an additional 20–30 minutes remain before radiation increases to hazardous levels.

What radiation doses are permissible? For the US Space Station doctors suggest that 600 rem should be the absolute maximum

Above: The Sun's vast, hot atmosphere is revealed in this X-ray view from Skylab.

Right: A solar flare bursts into space in this Skylab X-ray picture.

absorbed by the skin over an astronaut's career. For comparison: in August 1972 a particularly violent flare occurred. Had there been any unshielded space travellers outside the Earth's protective magnetosphere—the region dominated by the planet's magnetic field—they would have suffered a skin absorbed dosage of 2600 rem. This would be lethal. Even relatively benign flares can produce 100-rem skin absorbed dosages.

So shielding will be necessary on the Mars voyage. Perhaps the best option would be storm shelters surrounded by inflatable cylinders into which water could be pumped as needed. Even equipped with such shielding, a mission should be mounted during a period of low solar activity to minimize an astronaut's chance of contracting cancer.

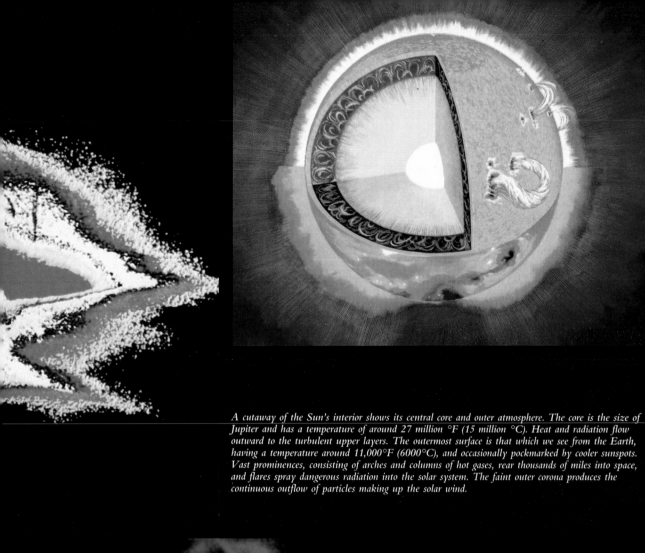

A cutaway of the Sun's interior shows its central core and outer atmosphere. The core is the size of Jupiter and has a temperature of around 27 million °F (15 million °C). Heat and radiation flow outward to the turbulent upper layers. The outermost surface is that which we see from the Earth, having a temperature around 11,000°F (6000°C), and occasionally pockmarked by cooler sunspots. Vast prominences, consisting of arches and columns of hot gases, rear thousands of miles into space, and flares spray dangerous radiation into the solar system. The faint outer corona produces the continuous outflow of particles making up the solar wind.

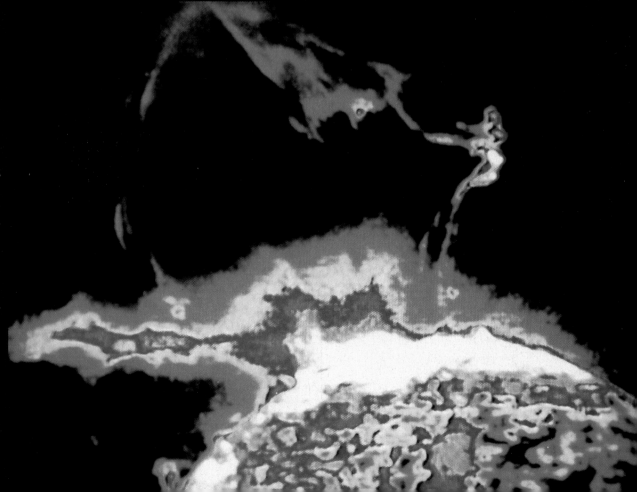

The workload on the crew and the ship's computers will increase rapidly as Mars expands into a globe and its gravity pulls the ship inward. Increasingly accurate trajectory data from observations made of the planet will confirm the position of the point of closest approach to within a few miles and reassure the crew that they are not heading straight for the planet.

The crew will need to carry out constant maintenance of equipment during the flight.

Flyby mission

There is a good chance that the first manned expedition will only be a flyby. That would mean that only a few hours of close-in flight would be available for a myriad of scientific experiments, so many of these would be preprogrammed for automatic execution at the busiest times.

Optimum launch windows occur every 15–17 years: the window in 2001 would be ideal for such a flyby expedition, and the next good one would fall in 2016. A 2001 launch would be on 9 March, flyby on 20 August (after a 172-day journey) and Earth return on 25 May 2002. The total mission duration would be 442 days.

It will be tempting to include an array of small probes to be ejected before the flyby: atmospheric descent capsules, penetrators and even orbiters. Penetrators could impact and carry out soil analysis

and seismometry and function as beacons for later landing missions. An orbiter would fire its braking motor at the point of closest approach, perhaps aiming to map potential future landing sites.

Mission planners will have decided what flyby distance the craft should aim at for the maximum scientific return. If it goes too close in, Mars will flash by too quickly. If it goes too far out, highly detailed views will not be obtained. The crew will be presented with a magnificent panorama: the planet will rapidly expand, spreading before them a landscape of canyons, craters and volcanoes.

Although a 2001 flyby would

automatically return the spacecraft to the vicinity of the Earth, it would have to increase its speed by 2860 miles per hour (4600 kilometres per hour) around the position of close approach to put it onto a precise path. It is even possible that a Venus flyby could be included in the homeward leg, but this would incur the cost of a 200-day increase in flight time.

Lander mission

On a landing mission the scientific observations would have to wait until the crucial manoeuvre of entering orbit around Mars had been accomplished. We assume below that rocket braking alone will be used; later, aerobraking is

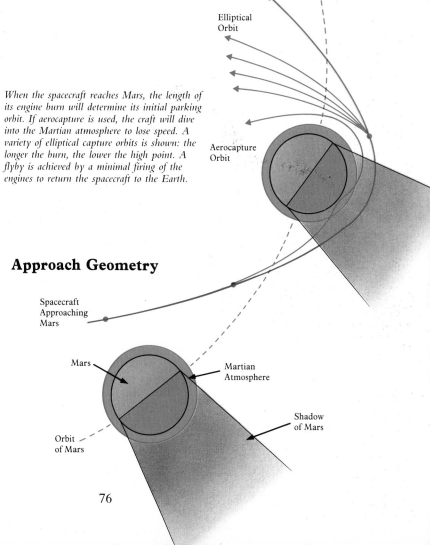

When the spacecraft reaches Mars, the length of its engine burn will determine its initial parking orbit. If aerocapture is used, the craft will dive into the Martian atmosphere to lose speed. A variety of elliptical capture orbits is shown: the longer the burn, the lower the high point. A flyby is achieved by a minimal firing of the engines to return the spacecraft to the Earth.

Fly-By

Elliptical Orbit

Aerocapture Orbit

Approach Geometry

Spacecraft Approaching Mars

Mars

Martian Atmosphere

Shadow of Mars

Orbit of Mars

described (See page 80).

As the ship approaches Mars, a computer will be programmed with information about when to ignite the engines and for how long. The burn will take place over the night hemisphere, because braking over the sunlit half would be inside the path of Mars and would involve reversing the spacecraft's motion around the Sun instead of bending it around the planet.

The crew will see the Sun slip below the Martian horizon and they will lose all contact with the Earth as the seconds to ignition tick by. They will be completely on their own as they skim above the dark landscape. As the computer's timer reaches zero, the engines will ignite and the crew will feel the entire vehicle shaking as it builds up to maximum thrust. They will be strapped in and feeling breathless from the unaccustomed forces following months of weightlessness. The prospect of the braking burn will have forced them to tighten up their housekeeping in the last hours and to tidy away or fasten down every loose item that might fly around during deceleration.

The timing of the burn is so important that a second computer and probably a third will be watching in case the first misses its cue; and as a last resort the pilot will have his own clock counting down

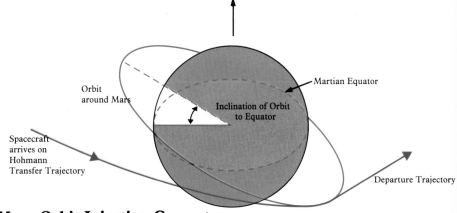

Mars Orbit Injection Geometry

The tilt of the spacecraft's orbit around Mars is determined by the plane change resulting from the Mars orbit insertion engine burn.

and will be ready to flick a manual ignition switch.

For several minutes the burn will light up the Martian night sky. Accelerometers will indicate the speed loss at every instant until the computers signal cutoff at a preset value. Again, if the computers fail, the pilot will be ready to step in to prevent an overlong firing from crashing them into the planet.

Though the computers will have predicted what the resulting orbit will be, the crew will check it by stellar tracking and discover whether they need to make any adjustments.

As the spacecraft emerges from behind Mars, the first bursts of telemetry will be transmitted; as much as 20 minutes later they will arrive at Earth and inform the helpless ground controllers whether all is well.

If the burn is late in starting, it might be possible to get into a looser orbit; but this might consume so much propellant that a landing would be ruled out and the crew would have to switch to a backup orbiting mission.

If the burn were to fail altogether, the spacecraft would sail past Mars. If it were not on a free Earth-return trajectory the engines would have to be fired as soon as possible to get it home. It might be possible to use the lander's engines to make up any shortfall in the homeward impulse from the main engine, just as Apollo 13 employed its lunar module descent engine when its main engine was damaged by an exploding oxygen tank. But once the craft is established in Mars orbit the engines *must* work again, or the crew are stranded.

Mission control will be vitally important to every aspect of the Mars voyage. Above: The Soviet control centre north of Moscow. Right: NASA mission control in Houston during the Apollo 13 flight.

The meaning of $\triangle v$

Engineers often talk about needing a particular '$\triangle v$' to carry out some space mission. The term comes from mathematics, where v is the standard symbol for velocity, and the Greek letter $\triangle$ (delta) represents a difference or change. So '$\triangle v$' means the change in velocity needed to carry out a particular mission. But what does this mean?

As an everyday example, consider a car driver approaching a hill. If he reaches the bottom at 60 miles per hour (97 kilometres per hour), turns his engine off and only just coasts to the top, then climbing the hill can be said to require a $\triangle v$ of 60 miles per hour. If the hill were higher, the $\triangle v$ requirement would be greater.

In interplanetary travel, $\triangle v$ is a measure of the energy required (and the difficulty) in getting from one place to another. To reach a given destination, space engineers have to calculate the $\triangle v$ required and then work out the size of vehicle, weight of propellants, etc, needed to get there.

To break free of the Earth altogether, a spacecraft has to attain 7 miles per second (11 kilometres per second): therefore, the craft needs a $\triangle v$ of 7 miles per second to leave the Earth.

Of course, merely to get into Earth orbit requires lower $\triangle v$s. Sputnik 1 had a $\triangle v$ of 4.8 miles per second (7.8 kilometres per second) or more. To reach NASA's Space Station in the 1990s, craft will need a $\triangle v$ of at least 5.7 miles per second (9.2 kilometres per second).

A table of $\triangle v$s is given here for the interplanetary bodies most accessible from the Earth. The figures reveal a remarkable fact of interplanetary mechanics: the moons of Mars are more accessible than our own Moon in terms of the total energy needed to land there. It should be pointed out that $\triangle v$s are cumulative: to go from the Earth's surface to the Moon, one adds the $\triangle v$s needed to escape from the Earth, to enter lunar orbit and then to descend to the surface. The main reason why the Martian moons are so accessible is that the energy needed to land on them is negligible—it could be provided by air jets. However, there are fewer launch opportunities for the moons and, obviously, the journey time is longer.

Because the orbit of Mars is significantly elliptical, $\triangle v$ require-

$\triangle v$ **Requirements for Interplanetary Targets**

Earth

Space Station in Low Earth Orbit (LEO).
$\triangle v$ = 9.2 km/sec

Phobos & Deimos
$\triangle v$ = 6.0 km/sec

Moon
$\triangle v$ = 8.3 km/sec

Mars
$\triangle v$ = 10.2 km/sec

$\triangle v$s FOR NEARBY TARGETS			$\triangle v$ in kilometres per second			
TARGET	LAUNCH FREQUENCY	TOTAL TRAVEL TIME	ESCAPE FROM L.E.O.	DIRECT LANDING	DIRECT RETURN TO EARTH	TOTAL (ASSUMING AEROBRAKING AT EARTH)
LUNAR SURFACE	FREQUENT	approx 7 DAYS	3.2	2.7	2.4	8.3
MARTIAN SURFACE (ATMOSPHERIC BRAKING)	EVERY 2 YEARS	approx 2 YEARS	3.6	approx 1.0	5.6	approx 10.2
PHOBOS AND DEIMOS	EVERY 2 YEARS	approx 2 YEARS	3.6	1.9	1.8	7.3
PHOBOS AND DEIMOS (MARS AEROBRAKE)	EVERY 2 YEARS	approx 2 YEARS	3.6	approx 0.5	approx 1.9	approx 6.0

Interplanetary Gravity Wells

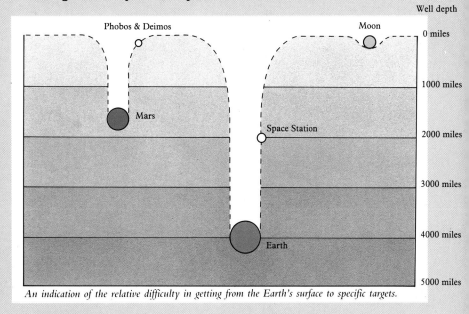

An indication of the relative difficulty in getting from the Earth's surface to specific targets.

ments vary appreciably from one launch window to the next. The minimum Δv required to carry a spacecraft from low Earth orbit to the vicinity of Mars is 2.2 miles per second (3.6 kilometres per second). This can be achieved only when the spacecraft flies a Hohmann ellipse meeting Mars at the planet's 'perihelion' (point of closest approach to the Sun). Journey time can be cut if Δv is increased, but that means sacrificing some scientific payload to accommodate more propellants.

Δv calculations are important for changing from one orbit to another. To change from the initial parking orbit to, say, a polar orbit is a difficult manoeuvre. The problem arises when going home: the spacecraft has to enter a parking orbit before it can transfer to Earth. The Δv requirements for a manned mission involving a polar orbit would require such vast amounts of propellant as to make it impracticable.

Engineers talk of a spacecraft having a 'Δv margin'—the maximum change in velocity that it can execute. This depends on the quantity of propellants it can carry. The Δv margin can be increased by the use of aerobraking or gravity assist. Voyager 2, for example, has flown by nearly all the outer planets with a Δv of 3.7 miles per second (6 kilometres per second): without using their gravitational influence, the journey would have needed a total Δv of 18.6 miles per second (30 kilometres per second).

Choosing Orbits

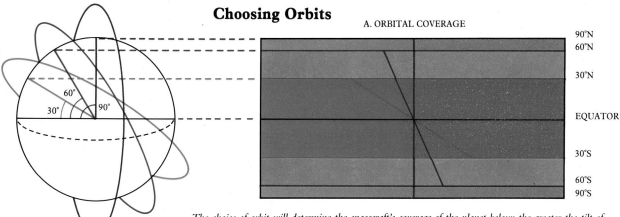

A. ORBITAL COVERAGE

_____ 30° INCLINATION
_____ 60° INCLINATION
_____ 90° INCLINATION

The choice of orbit will determine the spacecraft's coverage of the planet below: the greater the tilt of the orbit, the greater the area that will be covered. The simplest case is that of a circular orbit, for which the surface coverage depends on the orbital tilt, as shown (A), as does the track across the planet on successive orbital passes (B).

Choosing orbits

Although a polar orbit would be ideal for mapping landing sites, the energy, or Δv, required would probably preclude it (see box). It is more likely that the craft will enter an elliptical orbit with an inclination to the equator of about 30° at most.

One suggestion is that the craft

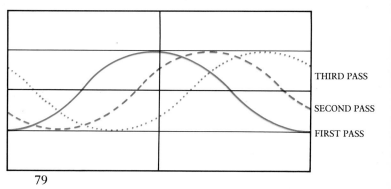

B. 'TRACK' ACROSS PLANET

Aerocapture

To date, spacecraft entering orbit around a planet have used their rocket engines to kill some of their approach speed in order to be captured by the planet's gravity. This manoeuvre uses up a great deal of fuel. For example, to brake into orbit around Mars in June 1976, the Viking 1 lander's main engine had to gobble up two thirds of its total fuel in a single 38-minute firing. If this retrofire phase could be avoided, the entire spacecraft could be made smaller and therefore cheaper, and its payload of scientific instruments could be increased. There is a way of doing this, although it has never yet been attempted.

In the 'aerocapture' or 'aerobraking' technique, a specially shaped vehicle would plunge into a planet's upper atmosphere, where it would lose speed through air friction before emerging into the desired orbit. Engineers consider it very difficult, and if adopted for a future Mars mission it will provide the most challenging design problems. If a manned aerocapture craft ploughs into the atmosphere too steeply, the frictional heating will burn through the protective thermal layers and incinerate the crew. If the entry path is too shallow, the craft will skip off the atmosphere like a flat stone skipping off water. Either of these could happen if the craft hits the atmosphere more than a mere one degree away from the perfect path.

Aerocapture has already been used in a small way. In 1968/9, when the unmanned Soviet lunar probes Zond 6 and Zond 7 returned to Earth, they skipped off

Principle of Aerocapture

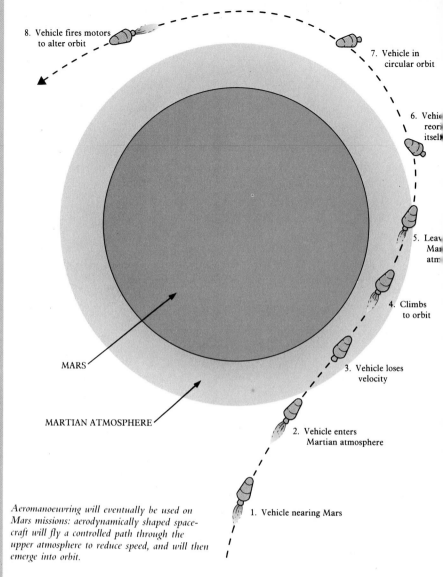

8. Vehicle fires motors to alter orbit

7. Vehicle in circular orbit

6. Vehicle reorients itself

5. Leaves Martian atmosphere

4. Climbs to orbit

3. Vehicle loses velocity

2. Vehicle enters Martian atmosphere

1. Vehicle nearing Mars

MARS

MARTIAN ATMOSPHERE

Aeromanoeuvring will eventually be used on Mars missions: aerodynamically shaped spacecraft will fly a controlled path through the upper atmosphere to reduce speed, and will then emerge into orbit.

Advantages of Aerobraking

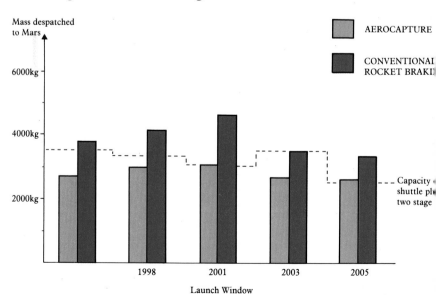

Mass despatched to Mars

AEROCAPTURE

CONVENTIONAL ROCKET BRAKING

Capacity of shuttle plus two stage

Launch Window: 1998, 2001, 2003, 2005

'These figures assume the rover vehicle enters an elliptical orbit with a perigee of 500km, and a period of 24 hours. Without aerobraking all but the 2003 launch window mission would not be possible.

the atmosphere briefly before full re-entry. This cut the peak deceleration level reached and stretched out the heating period of the craft.

It could be that aerocapture will not merely improve performance but make certain missions possible. NASA studies completed in January 1987 considered an unmanned aerocapture mission launched in the 1996 window. The orbiter was specified as weighing 1200 pounds (550 kilograms) and entering an orbit with a one-day period, with a low point of 310 miles (500 kilometres). A rover weighing 1410 pounds (640 kilograms) was to be landed. The study concluded that the mission could be accomplished with a 2.7-ton spacecraft launched by a combination of the Shuttle and an IUS (Inertial Upper Stage). A conventional rocket-braking vehicle would have to be a ton heavier and would require an additional kick motor.

The same study looked at all the launch windows from 1996 to 2005 and pointed out that there is little variation in the spacecraft mass required if aerocapture is used—the vehicle merely digs deeper into the Martian atmosphere to handle higher arrival speeds.

The penalty to be paid for the benefits of aerocapture is the increased complexity of techniques and systems. Basically the craft is aimed very precisely into the atmosphere and held on course until the guidance computer decides that enough speed has been lost. Entry speeds for all the launch windows from 1999 to 2028 range from 12,100 miles per hour (19,400 kilometres per hour) to 20,500 miles per hour (32,900 kilometres per hour). Speeds at the higher end could subject astronauts to 5.5g for several minutes. Doctors need to decide how much deceleration the astronauts can withstand for how long after months of weightlessness: Shuttle crews are limited to 3g after spending only 7–10 days in space.

should go into a highly elliptical 24-hour path for two to three days. At the low point, 220 miles (350 kilometres) high, subprobes would be dispatched into other orbits, some even to pass over the poles. Then there would be a burn giving a $\triangle v$ of 895 miles per hours (1440 kilometres per hour) at the low point to drop the high point to 12,610 miles (20,285 kilometres). This new orbit would take 13.5 hours to complete, and it would shift westward by 18° at every pass, so that in any 5.5-day period the craft would pass over all potential landing sites.

Aerobraking

Advanced missions will use aerobraking techniques to establish the desired orbit (see box). The gains from using this technique are certainly attractive, but it is difficult to carry out. An example of an approach sequence is the following.

The spacecraft has an initial speed of 15,970 miles per hour (25,690 kilometres per hour). It aims for a final orbit of $310 \times 20,450$ miles $(500 \times 32,900$

kilometres), which has a 24-hour period. The guidance computer directs the craft, which is aerodynamically shaped, for a closest approach of 28 miles (44 kilometres) above the surface—well within the atmosphere. At every point in its passage through the Martian air it constantly corrects the motion and calculates what orbit the craft would pop into if the computer should order full lift at that moment. This atmospheric foray lasts around six and a half minutes, during which time the crew experience a maximum deceleration of 2.4g (that is, they briefly 'weigh' 2.4 times their normal Earth weight).

If all goes well, the craft zips out of the atmosphere into an orbit with the desired high point but with a low point of 28 miles (44 kilometres). On reaching the high point, the craft fires its engines for a speed boost of 58 miles per hour (93 kilometres per hour), changing the shape of the orbit and lifting its low point to the desired 310 miles (500 kilometres). Next time round the orbit, the craft will avoid hitting the atmosphere.

Possible Orbit Around Mars For Manned Mission

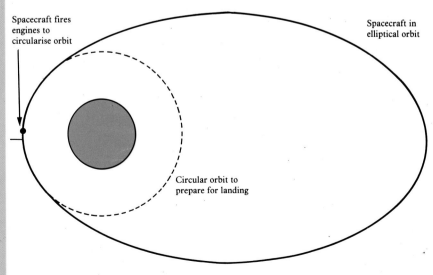

Spacecraft fires engines to circularise orbit

Spacecraft in elliptical orbit

Circular orbit to prepare for landing

A manned landing mission will probably enter an elliptical parking orbit, which it will then convert into a circular one in preparation for landing.

By the time the Viking orbiters ceased their operations, a total of 52,000 TV pictures of the Martian surface had been returned to Earth. Many were stereoscopic, some were in colour, and a few had a resolution sufficient to show objects on the surface that were as small as 80 feet (25 metres) across. Part of the Viking legacy is the confidence with which planners can now choose safe Martian landing sites for manned missions.

Man versus machine

The unmanned Soviet missions of the 1990s, together with NASA's Mars Observer, will further refine our picture of the Martian surface. But just how important are these 'precursors' to the landing of astronauts on the red planet? Some scientists suggest that they are vitally important; others believe that they are irrelevant. The latter cite the example of the unmanned Surveyor probes, designed to land on the Moon before the Apollo astronauts. The Surveyors are often regarded as precursors to the Apollo landings. Yet the Apollo lunar module's landing legs had already been designed before the first Surveyor reached the Moon in June 1966.

While it is true that a manned mission could be mounted with knowledge we already have, unmanned precursors would help considerably. Rovers, penetrators, balloons and orbiters equipped with very high-resolution TV cameras and radar could provide invaluable extra information to aid in the choice of target areas.

Choosing landing places

When the Viking mission was designed, target sites were chosen on the basis of Mariner 9 TV pictures. But when the Viking craft viewed these from orbit, they proved to be too rough. When new landing places were chosen, the scientific interest of the sites took second place to the landers' safety.

On future manned missions, safety will obviously be of even greater importance. But there will be greater flexibility in choice of landing site, owing to the human crew's ability to take over control of the landing, and the roving vehicles will provide extended range. This should ensure that sites of prime interest to the scientists will be within reach of the explorers.

The landing sites must be chosen on the basis of a global understanding of the surface, rather than by the attractiveness of specific localized features, however intriguing they may be. The selection must take into account:

- the difficulty of guaranteeing that the lander will touch down at a specified place;
- the proximity of interesting geological features;
- the probability of locating recently exposed rocks;
- the planned range of the roving vehicles.

In regions where there are extensive dunefields and little visible bedrock, surface mobilities of a few miles may be necessary to ensure that rock samples can be obtained. Naturally, landing sites that offer access to more than one interesting type of terrain have clear advantages. Unfortunately, such sites tend to be rough and may pose significant hazards and obstacles to roving vehicles.

Key regions

The geology of the planet is covered in greater detail in Part 4 of this book: what follows is merely an outline.

NASA Surveyor landers preceded the first manned lunar landings.

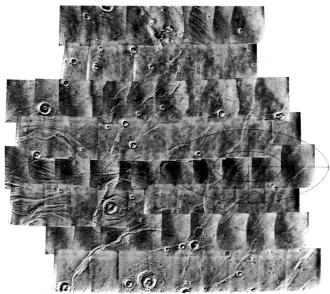
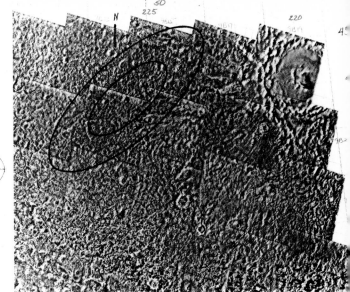

The landing sites for Viking 1 (left) and Viking 2 (right) were chosen for their relative safety. Imaginary ellipses were drawn around the sites: there was a 99 per cent probability that the Viking would land within the larger ellipse, and a 50 per cent chance that it would land within the smaller.

Simplified Geological Map of Mars

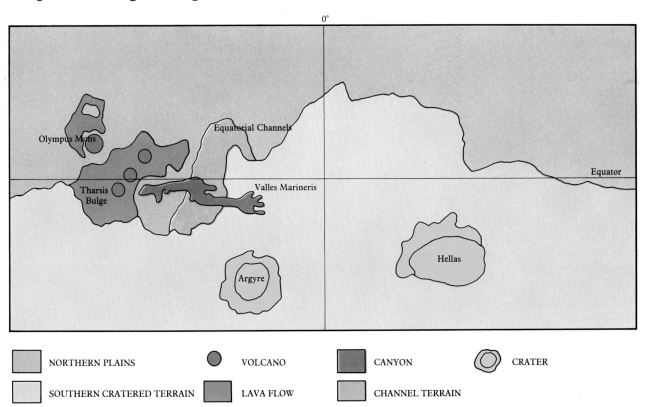

The choice of landing site will be verified by crew observations from Mars orbit.

- young volcanoes;
- middle-aged volcanoes;
- ancient cratered terrains;
- layered regions;
- polar regions.

The candidate sites

On this basis, 10 prime landing sites have been identified for manned missions by specialist NASA committees: see table. The Soviets have identified seven similar sites, of which the primary ones are at the eastern edge of Kasei Vallis and near the volcano Uranius Patera.

The Kasei and Mangala Valles are made up of channels apparently carved by flowing water in the remote past. Detailed studies would provide important clues to one of the most intriguing questions of Martian planetary science: how Mars lost its primordial reserves of water. These locations offer the bonus that they are close to the large equatorial volcanoes and volcanic plains, which will have affected their geology significantly.

There are a wealth of interesting sites other than these, but they present difficulties for an initial manned mission. Polar sites are precluded for manned missions: it is costly in fuel to change from the initial parking orbit to a tilted orbit in order to land there. Among non-polar sites, it would be very difficult to land on the gigantic volcano Olympus Mons or in the huge Valles Marineris canyons. Southern hemisphere sites are less attractive than those in the northern hemisphere, because they are rougher. And such sites are closer to the areas where the regular dust storms have their origin, which might be inadvisable.

But these areas will not be left undisturbed by visitors from the Earth. In due course they could all be visited by long-range vehicles, manned or unmanned, sent from bases elsewhere.

The surface of Mars shows a remarkable diversity of features. There is a striking difference between the two hemispheres of the planet. The southern is heavily cratered and shows ancient surface terrain; the northern is made up of smoother, younger plains. Four main geological 'provinces' have been identified.

Ancient regions are composed of heavily cratered terrain and mountains. The rocks here would provide information on the early Martian crust, enabling the time of its formation to be determined.

Volcanic regions include the great equatorial volcanoes and the northern volcanic plains. It is likely that rock samples in this region will reveal lavas of differing ages and chemical compositions, representing different chapters in the geological history of Mars. They may reveal the role played by volcanic gases in forming the early Martian atmosphere.

Polar regions are composed of permanent ice, layered deposits and etched plains. The rocks in this region are expected to reveal sedimentary strata that could provide important records of the history of interactions between the Martian atmosphere and the surface. There may be water ice in the polar subsoil, which means these are important areas in which to search for traces of life.

Modified regions are areas where crustal processes have altered terrain of the other three types. These processes may have brought to the surface material that is inaccessible elsewhere. Samples of material taken close to relatively young impact craters may contain subsurface material that has not had time to be modified by exposure to the atmosphere.

Based on this current understanding of Mars, the key geologi-

Proposed manned landing sites on Mars

GEOLOGICAL TYPE	CANDIDATE SITE	REASONS FOR CHOICE	PROBLEMS WITH SITE
Young volcano	Olympus Mons (18°N, 133°W)	The youngest and tallest Martian volcano; youngest recognized lava flows emanate from it.	Elevation of 17 miles (27km) is so high that sufficient aerobraking could not be achieved.
Middle-aged volcanoes	West of Olympus Mons (20°N, 150°W)	Area close to middle-aged volcanoes, covered with their lava flows.	These volcanoes are as high as Olympus and cause the same problem with aerobraking.
Ancient cratered terrains	Argyre Basin (49°S, 43°W) Isidis Basin (15°N, 270°W)	Both sites are large, multi-ringed impact crater basins, among the most ancient craters on Mars.	The multiple rings of the basins are mountain chains, dangerous to land on or to negotiate.
Layered regions	Candor Chasma (6°S, 75°W)	Exhibits layered terrain in walls of the Valles Marineris canyon system, providing clues to the canyons' origins.	A complex site that would require many weeks of systematic study, traversing difficult terrain— 2.5 miles (4km) deep in places.
Polar regions	Chasma Boreale (85°N, 110°W)	Region between residual ice cap and layered terrain, important for climatic studies.	Difficult site to reach because of high latitude.
Hemisphere boundary	Mangala Vallis (10°S, 150°W)	Equatorial site dissected by ancient channels.	Perhaps the most accessible site.
Equatorial channels	Kasei Vallis (21°N, 80°W)	Channels created by catastrophic flooding in past.	To gain a systematic picture long traverses would be needed.
Ancient crust	Nilosyrtis Mensa (32°N, 290°W)	'Fretted' terrain, showing possible sections of ancient Martian crust.	Difficult terrain to negotiate; again, long traverses needed for systematic surveys.
Ancient volcano	Amphitrites Patera (60°S, 300°W)	Evidence for ancient volcanism, with eruption of ashlike material.	Difficult site to reach because of high latitude; limited scientific objectives.

Possible Manned Landing Sites

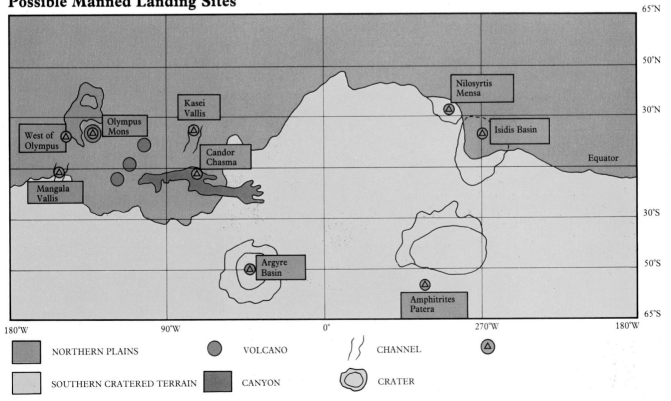

NORTHERN PLAINS   VOLCANO   CHANNEL

SOUTHERN CRATERED TERRAIN   CANYON   CRATER

THE DESCENT

In whichever way it is done, dropping from orbit onto the Martian surface is the most hazardous part of the mission. It involves a manned spacecraft slamming into the atmosphere of a distant, hostile planet at 9000 miles per hour (15,000 kilometres per hour). The landing site itself, however carefully chosen, could harbour dangers unsuspected until the last moments of the descent.

When the preparation for landing begins, the lander is powered up and checked out in detail by the crew, working from a checklist and helped by onboard computers. Help from mission control on Earth is limited by the delay in radio communications. Each system is activated in turn and faults are, if possible, repaired. The computers themselves are thoroughly tested: the mission rules will almost certainly demand perfect operation before lander separation. Power supplies, communication links, life-support systems, propellant levels and tank pressures, and valve settings are all vetted. The small attitude-control thrusters may be 'blipped' to give confidence that they will again function satisfactorily.

Lander separation

After the lander checkout has been completed and the entire crew has enjoyed a full sleep period, they put on their full pressure suits. The explorer crew crawl through to the lander. The hatches are closed and airtightness tested before everyone straps in for separation. The docking latches are released and the residual air pressure in the transfer tunnel gently pushes the lander away. Brief firings of its thrusters move the lander to a safe standoff distance, from which the crew in the mother ship can inspect it as it slowly revolves.

What happens next depends very much on the spacecraft design and the choice of parking orbit. If the vehicles are circling, say, 220 miles (350 kilometres) above Mars, the lander could make a direct descent. The deceleration involved would be 2–3g.

However, the two craft are more likely to be in a highly elliptical 24-hour parking orbit with low and high points of 310 miles (500 kilometres) and 20,450 miles (32,900 kilometres). If a spacecraft braked from this orbit in the manner of an Apollo CM (command module) returning to Earth, the

crew would suffer a peak deceleration of 4–5g. After months of weightlessness this could be severely debilitating. So it is likely that the lander, or an orbital tug connected to it, will fire braking engines at the low point of the lander's orbit to pull down the high point.

In the lander's new orbit, the crew and computers track the main ship and continue to monitor the lander's systems in case they have to make a premature return. In the event of a major failure the main craft might have sufficient propellant to swoop down for a rescue, but it needs to conserve an adequate margin for the large Earthbound burn. A separate orbital tug could be valuable here, since it could carry out such a rescue at a lower fuel cost.

Once crew and ground control are satisfied that the spacecraft is properly established in its orbit and is in full working order, the lander is turned to fire its braking rockets. This lowers its trajectory so that it will enter the atmosphere. It then rotates again so that its heatshield faces the direction of entry. Mars looms larger and the landscape below slips by faster as the lander descends towards atmospheric entry at around 30 miles (50 kilometres) above the surface.

At first the crew, able to see only the blackness of space through their windows, see little sign of their entry. The first indication might be a '0.05g' light, of the kind carried on American spacecraft to indicate that atmospheric drag is beginning to bite. Deceleration builds up to the precalculated peak and a glowing sheath of ionized air

During the descent through the Martian atmosphere the cabin will be suffused with a pink glow from the hot gases enveloping the craft, rather like this scene in a Space Shuttle during re-entry.

APOLLO TYPE

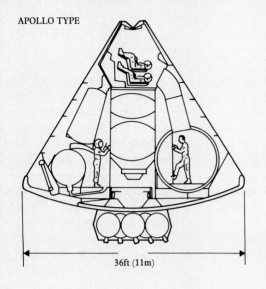

36ft (11m)

LARGE AEROSHELL

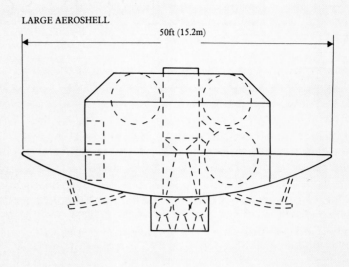

50ft (15.2m)

AERODYNAMIC

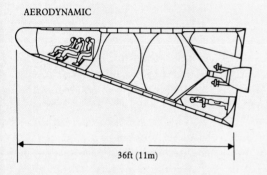

36ft (11m)

WINGED AERODYNAMIC

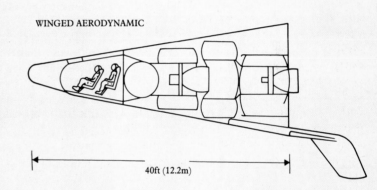

40ft (12.2m)

A range of American Mars lander designs. Top left: An Apollo-type vehicle based on the conical command module designed to plunge through the atmosphere heatshield-first. Its shape limits its manoeuvrability and hence the choice of landing site. Its advantage is its simplicity: it could even be landed manually if the computers broke down. And development would be relatively simple and low in risk. These factors make it a likely candidate for an early American landing mission—for example, the proposed Sprint project. Top right: a cargo lander with a Large Aeroshell that is jettisoned after braking through the atmosphere, clearing the landing legs for release. The lower engines are then fired to brake out of orbit and are discarded before the craft hits the atmosphere. A variety of supply modules are protected behind the large shell.

The lower pictures show two more sophisticated Aerodynamically-shaped lander types that could be introduced on later missions. Their aerodynamic 'biconic' body shapes and (bottom right) wings make them far more manoeuvrable than the Apollo-type lander, but no spacecraft like them has ever been built. Their guidance and control systems would be even more complex than those of the Space Shuttle, which cannot be flown without computers. Their angled entry also dictates heat shielding over most of their bodies. They would touch down base-first, for a vertical lift-off at the end of exploration.

The simpler Apollo-type craft could probably not land directly from the Earth–Mars approach trajectory because the deceleration would be too great for human crews. They could be used only on missions in which they could first brake into Mars orbit. The aerodynamic craft could make direct landings, because they would bleed off speed gradually by skimming through the atmosphere.

envelops the craft as the temperature rises, cutting off communications with the mother ship. When Viking 1 descended on Mars, the surrounding air was heated to a peak of 2700°F (1500°C) and its deceleration reached 8.4g. This value is far too high for a human crew.

Ready to escape

The descent is under computer control for the most part, but the crew would be able to take over if needed. In a dire emergency they could punch the abort button to get back into orbit. All the way down, the computer knows at every instant what has to be done

to reach orbit again. If the lander is Apollo CM-type, the crew is strapped inside the return stage, so in the event of an abort the ascent engine would fire. If, on the other hand, the craft is an aero-manoeuvring biconic (see box), the computer might be able to use its lift combined with a boost from its small thrusters if the escape is made while still high in the atmosphere; from a lower height the main engine could be fired.

If all goes well, the computer continuously predicts a touchdown area based on the flight path up to that point. The biconic craft's high lift permits it to make extensive manoeuvres in the atmosphere, so (like the Space Shuttle) it has a wide choice of landing sites. An Apollo CM-type craft is more limited.

The lander's computers will be processing a stream of data from external pressure sensors, accelerometers (to measure the rate of slowdown), gyros (to sense the craft's attitude), and, towards the end, the ground radar. The radar's speed and height information is vital. The Apollo 14 mission almost had to abort when its radar failed

to lock onto the surface at the intended height of 7.5 miles (12 kilometres); in fact it locked on at 4.3 miles (6.9 kilometres).

The lander may receive navigational assistance from radio beacons already in place on the surface. The Soviets have talked of firing such beacons into the ground inside penetrators, to provide guidance for their unmanned rovers in the 1990s. They would function rather like airport landing aids, providing distance, height and speed data.

Apollo showed that a CM-type lander can make a manual entry if a computer should fail, with the pilot rolling the spacecraft to vary lift. An aeromanoeuvring vehicle, however, would be trickier to land—the Space Shuttle cannot be flown completely manually—so in such a case it might have to abort.

Human backup
The crew do not have a direct view of the surface. All they can see through their windows is the

The final stages of powered descent are shown below. The crewed lander vehicles are descending towards a prearranged landing site where cargo vehicles have already landed safely. The crew vehicles remain upright after landing, ready for any emergency return to orbit.

sky, becoming brighter and pinker as they descend. But downward-viewing high-resolution colour television, possibly stereoscopic, helps to guide them. Whatever the type of craft they are flying, the astronauts will probably take over in the final stages because human beings are the best judges of a landing site's suitability. (Neil Armstrong took control during the first manned Moon landing when he saw that the computer was directing the lunar module *Eagle* into a rough area.) Of course, the landing attempt is made in daylight, and probably during the target area's mid-morning, when the angle of the Sun gives the greatest surface contrast and there are still plenty of daylight hours left after touchdown.

Whereas an aeromanoeuvring vehicle can fly a convoluted path to bleed off speed, an Apollo CM-type has to rely on other methods. A balloon/parachute device (a 'ballute') could inflate about six miles (10 kilometres) up to slow the descent rate below supersonic speed in preparation for the main parachutes to pop out. These would be much larger than their terrestrial counterparts because the Martian atmosphere is so thin. When the parachutes have been deployed the heatshield can be released and the landing legs opened out. Parachutes are jettisoned and the soft-landing retro-rockets fired about half a mile (one kilometre) high, some 15 minutes after entering the atmosphere.

Touchdown is at no more than six miles per hour (10 kilometres per hour). Before the congratulations can begin the crew must immediately configure the craft for a possible abort. The mother ship's orbit will probably have been arranged so that contact is maintained throughout descent, apart from the brief blackout period. But it will be another 20 minutes or so before radio signals reach ground control and tell the Earth of the lander's fate.

Mars descent profile: The lander separates from the orbiter, which continues to circle the planet. The lander orients itself to fire its engine for a deorbit, and enters the Martian atmosphere about 30 miles (50km) above the surface. The lander experiences frictional heating for a few minutes; then, to lose speed, it either deploys a ballute or aeromanoeuvres. At around 5 miles (8km), the lander deploys a drogue to open its main chute. It discards its heat shield and deploys its legs. The craft is now guided by its onboard radar, locked onto the Martian surface, and by data from radio beacons on the surface. The main chutes are jettisoned, and the retro-rockets are fired to slow the descent speed further. During the final descent, the retro-rockets continue to fire to achieve a touchdown within the planned landing ellipse.

VIEW FROM THE SURFACE

When the Viking landers returned pictures of the Martian surface, there was some surprise that the landscapes appeared very similar to each other, for the landing sites had been chosen on the basis that they had different geological features. The lander pictures showed striking panoramas of rock-strewn dusty desert, coloured a vivid orange by 'rust'—iron oxides. Planetary scientists believe that the Viking views are largely representative of the Martian surface.

Rocks and dust

Viking 1 landed only 20 feet (8 metres) from a large, dark boulder later named 'Big Joe'. It was about 3 feet (1 metre) high and 9 feet (3 metres) across, and it would have destroyed the lander if touchdown had been closer. One of Viking 2's footpads lodged on a small rock, which tilted the craft and gave the horizon an apparent eight-degree slope.

Many of the rocks at the second site are pitted by small holes, giving them something of the appearance of Swiss cheese. Probably the rocks are volcanic in origin and the holes were formed by gases escaping while the rocks were molten. Also visible in many Viking pictures are outcrops of rocks that have been carved into odd shapes by extensive wind erosion.

The surface soil has much the same physical properties as sand on a beach where a wave has just washed up, though it is actually much finer-grained than sand. The trenches dug by the Viking landers remained visible for many months. Chemical analysis of the soil excavated from those trenches revealed a composition similar to that of basaltic rock on Earth.

What look like sand dunes are also visible in many Viking lander views. Geologists prefer to class them as drifts that persist for long periods, as opposed to the dunes of terrestrial deserts, which are transient. Viking scientists were surprised that as time passed the lander cameras recorded only minor changes on the surface, despite violent winds and dust storms.

Dust storms

Perhaps the biggest surprise was the brightness of the sky and its colour: pink. The thin atmosphere carries dust particles that scatter the red light more effectively than other wavelengths.

Almost every Martian year, there are giant dust storms, which can engulf the whole of the planet. The dust particles, no more than a few hundred-thousandths of an inch (a thousandth of a millimetre) across, are whipped up to altitudes of 27 miles (45 kilometres) by strong winds. It takes months for the dust to settle, and the effect at the surface is to gradually diminish the strength of sunlight.

The global dust storms could cause problems for the first astronauts on Mars, so the first landings will be timed to avoid them.

Clouds and frosts

Mars is an arid, frigid planet with temperatures and pressures so low that water can exist only as ice or vapour. Though the planet may have been wetter in the past, only

A few minutes after touchdown in September 1976, Viking 2 returned striking views of pitted, volcanic rocks close to one of its footpads (left). A colour view returned a few weeks later (above) has been enhanced to reveal subtle colour differences in broken nearby rocks.

Below: A panorama of the rock-strewn Martian surface, taken from the second Viking lander. The first lander narrowly avoided a large rock (top right), christened 'Big Joe', whose dust-covered appearance changed little for the entire period during which it was observed. Dust drifts, looking like terrestrial sand dunes, are visible in the foreground: distinct ripples were also seen in them (left). The passage of a global dust storm in 1980/81 is seen in the sequence above (top left). To enhance the dust storm (third from left) the colours have been altered, making the sky appear artificially blue; however, the lighting levels are correct. (One 'sol' is a Martian day, slightly longer than our own.)

small amounts of water remain, mostly locked up in the subsurface permafrost.

Nevertheless, explorers may sometimes see thin wisps of cloud in the Martian sky, if these are not masked by the pink dust haze. In middle latitudes they consist of water ice, but nearer the poles they are made up of carbon dioxide crystals.

During winter, frost also settles on the surface at higher latitudes. Seen just after sunrise it is probably composed of a mixture of water and carbon dioxide ice crystals that condense on the surface dust. The Martian atmosphere is so cold and dry in winter that it cannot hold enough water vapour to produce water-ice frost on its own.

Dawn on Mars

Sunrise on Mars will be fascinating for the first explorers. The familiar constellations will fade from the sky as a tinge of pink rises from the eastern horizon to spread across the whole sky. The Sun that rises above the horizon will be a feeble version of the one we see from Earth: less than half as bright, and only two thirds the diameter.

There will be something comfortingly familiar about the pace at which the Sun crosses the sky. Mars rotates on its axis in a period that by pure coincidence is only 40 minutes longer than the

A thin layer of surface frost, composed of water and carbon dioxide ice, was recorded by the Viking 2 lander during winter.

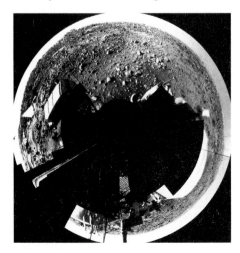

A computer-generated 'fisheye' view of the Viking 1 landing site reveals a rock-strewn desertscape.

Earth's day (though this will still cause problems with timekeeping on the Martian surface).

On Mars the height of the Sun at noon varies with the time of year, as it does on Earth. Again by pure coincidence, the planet's axis is tilted by almost exactly the same amount as the Earth's—about 24°—and so it experiences similar seasons as it goes around the Sun. Because Mars's orbit is elliptical the lengths of the seasons vary.

A sky with two moons

Near the equator the sunshine is dimmed briefly every few hours as a black shadow crosses the face of the Sun. This is an eclipse of the Sun by Phobos. While the Earth's Moon totally hides the Sun during an eclipse, Phobos, as seen from

the surface of Mars, blocks off only half the Sun's disc. When Deimos, the smaller moon, passes in front of the Sun, it appears only as a black speck against it.

After sunset the two moons come into their own. On Earth we are used to seeing the phases of the Moon change progressively during the course of a month. On Mars, each moon's cycle of phases is speeded up by comparison.

The smaller moon, Deimos, goes through its cycle of phases in only 30 hours—little more than a Martian day. If Deimos is full when it rises in the east, the Mars explorer is in for a fascinating show. Deimos appears to move across the sky, as the stars do, because Mars is rotating. But all the time its own motion is taking Deimos the other way in relation

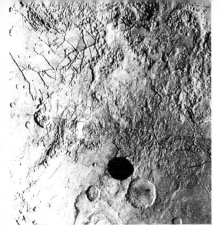

The Viking 1 orbiter viewed Phobos against the background of the planet.

to the stars, like someone walking up a down escalator. In the course of a night it is left behind by the stars and moves less than a fifth of the way across the sky. It is still in the eastern sky when the Sun rises, and it has waned visibly to become a mere crescent.

Unfortunately, Deimos is not only the smaller of the Martian moons, but also the farther from the planet. As a result it appears tiny (only a fifteenth of the diameter of our Moon as it appears to us), and it is difficult to make out the phases with the naked eye. In fact Deimos looks like a brilliant star, a few times brighter than Venus looks to us.

Phobos is another story altogether. It shines 40 times more brightly than Deimos, and is both

large enough and close enough for the Mars explorers to see its shape—a slightly oval moon with almost two-thirds the diameter our Moon appears to have to us. Phobos is so close to Mars that it whirls around the planet in less than eight hours—far faster than the planet itself rotates. This means that Phobos is like someone running up an escalator faster than the escalator is moving down. So it rises in the west and sets in the east.

The Mars explorers can see the phases of Phobos easily, because when it is right overhead it looks 40 per cent as wide as our Moon looks from Earth. But they can never see a full Phobos: when it is opposite the Sun in the Martian sky, it is in the planet's shadow, suffering a total eclipse.

The night sky

When the stars come out in the Martian night, the constellations look just as they do from the Earth. The journey from Earth to Mars is far too short to affect our perspective on the stars. But the constellations do seem to lie at an odd orientation. The reason is that the axis of Mars points in a different direction in space.

The brightest objects in the Martian night sky are the other planets, as they are in the Earth's. But the appearance of these has certainly changed. The outer planets look more impressive than they do from Earth; Jupiter, for example, looks 30 per cent brighter. But it's the inner planets that have changed the most.

The planets within Mars's orbit never move far from the Sun, and so shine as morning or evening stars. Mercury is even more elusive than it appears from the Earth: dim and always very near the Sun. Venus is farther from Mars than it is from the Earth, so it does not have quite its usual brilliance.

But there is a third, unfamiliar, 'star' that appears in the morning or evening skies of Mars. It is a bluish planet, almost as bright as Venus, straying further out from the Sun's glare. Someone with good eyesight may see that it is not alone: close to it and going round it every 27 days is a dim brownish object, only one-hundredth as bright. This beautiful azure star, hanging in the pink skies of a Martian sunset or dawn, is the world the travellers have come from—planet Earth, in company with its Moon.

Sunset on Mars as recorded by the first Viking lander.

SETTING UP BASE

Once the newly arrived explorers are satisfied that no emergency is imminent they can begin to power down various systems in order to conserve their limited energy supplies. Only life-support systems, communications links and essential services will be kept functioning. One computer might be left on to monitor the craft and be ready to handle a hasty departure.

The Apollo lunar modules permitted only three-day stays on the Moon, and their cramped interiors provided only the most basic facilities: there weren't even any seats. Even an initial Mars landing, with a crew of two or three men staying for two to three months, must provide a few more of the comforts of home.

Even so, the lander will be cramped since it is unlikely to have separate modules for habitation, for scientific equipment and for life-support systems. It will lack the luxury of a shower—damp cloths will have to suffice. The life-support systems will have to meet the basic daily requirements of each man: 1.84 pounds (0.83 kilograms) of oxygen, 1.6 pounds (0.73 kilograms) of solid food, 4.1 pounds (1.86 kilograms) of drinking water, 1.6 pounds (0.73 kilograms) of water for food preparation, and the removal of 2.2 pounds (1 kilogram) of exhaled carbon dioxide and 3.3 pounds (1.5 kilograms) of urine.

Power supplies

Six fuel cells, which generate electricity from the chemical combination of hydrogen and oxygen to form water, could provide sufficient power, warmth and water for a 60-day stay. Such an 'open-cycle' system does not recycle waste water and exhaled carbon dioxide. This creates the problem of what to do with the waste: quarantine rules might forbid

dumping it on Mars. So the first Mars mission might introduce some elements of a more sophisticated 'closed-cycle' system. Thus water could be broken down electrically to produce oxygen for breathing, as on the Mir space station.

It could be that small nuclear power packs, like those that have been used in several unmanned deep-space probes, would be used instead of fuel cells. Solar panels would be safer, but the strength of the sunlight on Mars is only two fifths of that on the Earth, and output from solar cells would cease at dusk.

In principle, oxygen, hydrogen, water, nitrogen (for fertilizer) and methane (for fuel) could be extracted from the thin Martian air. Any subsurface water permafrost that might be found could be 'mined'. These processes might be important for future long stays and permanent bases, but the equipment required would be too heavy for the first trip, and at most the principles could be demonstrated on a small scale.

The first Marswalk

Once the go-ahead for a long stay has been given, pressure suits will be taken off with relief. There will be a meal and a night's sleep before the first venture onto the surface. But the necessary chores following a Mars landing are likely to be more protracted and tiring, and night will soon arrive (whereas on the Moon there are 14 days of daylight).

When the explorers put on their suits again, they will add a backpack for independent movement. The suits will be more comfortable than the Apollo Moonsuits, and more easily maintained and replenished. The Apollo backpacks had an Earth weight of 200 pounds (91 kilograms), but a Moon weight of 33 pounds (15 kilograms). They supported seven hours of EVA (extravehicular activity). On Mars there will have to be a lighter version, which will therefore be able to support only two to four hours of EVA—a strong constraint on surface exploration. The backpack will provide not only oxygen and power, but also small amounts of food and water, supplied through tubes inside the helmet. The Apollo Moonsuit cooled its occupant by boiling water off into the surrounding vacuum, but Mars is so much colder that a radiator arrangement might be sufficient.

The cabin's atmosphere will be a mixture of nitrogen and oxygen at a pressure of about 0.7 atmospheres. For engineering simplicity the Mars suits will use pure oxygen at a lower pressure (see box). If an explorer donned a suit and went straight out of the ship, he would suffer an attack of the bends—bubbles of nitrogen forming in his bloodstream, leading to painful and possibly fatal cramps. To avoid

Preparing for a Marswalk. The astronaut first dons an undergarment threaded with tubes linked to his backpack's water-cooling system.

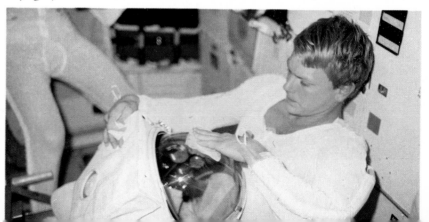

Viking 1's sterilized protective aeroshell (left) gouged out a small crater, half a mile (750 metres) north of the lander site itself (small arrow). The crater (large arrow) was not discovered until orbiter pictures were analysed in 1987.

Contaminating Mars

The question of the existence of life on Mars is still controversial. Until samples returned by manned or unmanned missions have been thoroughly studied, strenuous efforts will be made to avoid contamination by Earth bacteria. When solar system exploration began in earnest in the 1960s, the international scientific community became concerned that terrestrial organisms might establish themselves on other planets. Mars was of particular concern because, of all the Earth's neighbours in the solar system, it was the most likely to harbour life of its own. A 'dirty' spacecraft, searching for life, would detect the Earth organisms that it had brought along with it, and these would also confuse the picture for later missions. Mars is a planet-sized laboratory for exobiologists (scientists who study life beyond the Earth) and they do not want it to be marred.

COSPAR, the International Committee on Space Research, first met in 1958 to discuss the problem, and eventually decided that standards should be set such that there would be less than one chance in 1000 of contaminating Mars for 50 years, the period of 'biological exploration'.

Little is known of Soviet sterilization procedures, but great care was taken by the United States to minimize the risk that the twin Viking landers of 1976 would con-

taminate the planet. Each was encapsulated in a 'bioshell' and heat-sterilized at 234°F (112°C) for 40 hours. The bioshell prevented subsequent recontamination of the craft.

Viking's search for life was inconclusive, but it boosted confidence that terrestrial organisms could not multiply on Mars. Harsh ultraviolet radiation bathes the surface, and the soil is harmful to life because it is highly reactive chemically. If it turns out that nitrogen does not exist in the soil, that factor alone would limit the potential survivors of terrestrial organisms to blue-green algae and bacteria.

Nonetheless, there is no room

Earth bacteria might have survived 31 months on the Surveyor 3 Moon lander.

for complacency. When Apollo 12 astronauts brought back pieces of the old Surveyor 3 Moon lander, biologists found evidence suggesting that Earth bacteria had survived on it for the 31 months since it had left the Earth (though it is not certain that the material was not contaminated some time after collection).

Orbiters are subject to less stringent requirements. The rules are that there must be no more than a 1 in 10,000 chance of impact before 31 December 2008 and then at least a 95 per cent probability of remaining in orbit until the end of 2038. The Soviet Phobos spacecraft will be quite safe in their high orbits (see page 20); but the American Mars Observer will have to be raised at the end of its mission from its 224-mile (360-kilometre) orbit to one at 310 miles (500 kilometres) to avoid atmospheric drag.

Contamination from a manned landing is far more likely and possibly even unavoidable in the short term. Before a manned expedition is mounted the Martian soil must be studied to discover which terrestrial organisms can survive and under what conditions. If it turns out that some of them can flourish on Mars, mission engineers will be faced with the enormous challenge of completely sealing spacecraft, EVA suits and manned rovers to avoid any leakage into the Martian biosphere.

this, Marswalkers will have to breathe pure oxygen for several hours before exit to wash the nitrogen out of their blood.

If there were no airlock, all the air in the main cabin would have to be pumped out to avoid losing it when the hatch was opened. Providing an airlock leads to less wastage of air, though it increases the weight of the craft. And it permits one astronaut to remain inside the craft to monitor the progress of the EVA, without the inconvenience of getting suited up.

One small step . . .

As the walkers emerge into the sunlight, they will lower visors coated with a thin, transparent layer of gold to cut out dangerous ultraviolet radiation. The honour of being the first down the ladder and into the history books will probably go to the commander, who will doubtless be primed with a suitable historic phrase to match Neil Armstrong's 'That's one small step for a man, one giant leap for mankind'.

The walkers will inspect their craft for visible signs of damage and will grab 'contingency' soil samples in case they have to make a hasty retreat. At some stage they will unload a roving vehicle. The Apollo Lunar Roving Vehicle had an Earth weight of 460 pounds (209 kilograms), but Martian gravity, greater than the Moon's, will demand greater robustness,

When man takes his historic first steps on Mars, the scene could well resemble this 1971 Apollo view.

and the Mars rover will be needed for a longer period. So it will probably have an Earth weight of around 1100 pounds (500 kilograms), though on Mars this will be only 430 pounds (195 kilograms). If it is powered by batteries, they will be rechargeable from the lander. Its use will be limited to around three miles (five kilometres), because the astronauts need to be able to walk back if it breaks down. It will also be possible, should one backpack fail, to connect the suit by umbilicals to another, or to the rover.

Within the limits of the rover's range, the astronauts will take soil samples as far removed as possible from the lander's contamination. They will gather several hundred pounds of surface soil, rocks and 'cores', drilled from beneath the surface. They will take numerous 30-foot (10-metre) cores, and at least one 300 feet (100 metres) long, which will be studied in the lander and then broken up so that parts of it can be returned to Earth.

When they have completed their EVA, the explorers will re-enter the lander's airlock, with their samples sealed inside boxes to be preserved in pristine condition. The astronauts will vacuum-clean each other and all their equipment to avoid carrying dust into the cabin. The Martian surface dust, as detected by Viking landers, is very highly oxidizing. If it were to be inhaled, it could do untold damage to astronauts' lungs.

Even though each EVA might last only two to four hours, preparations and cleaning up will make a very full day. If there is a three-man crew, different pairs could make successive forays in rotation, a system that worked well on NASA's Skylab. Earth control will monitor the astronauts' biomedical condition and watch out for early signs of exhaustion. Days off will be essential, even though everyone will be eager to get the most out of the short time available.

Dressed for the occasion

Since a Mars mission will be lengthy, with many EVAs (extravehicular activities), the spacesuits worn on the surface need to be flexible, comfortable for long periods, reliable, and easy to maintain and replenish.

A spacesuit can be made more reliable the lower the pressure of the air it contains. Accordingly, the Mars suit will supply 0.2 pound (0.1 kilogram) of pure oxygen each hour, at about a quarter of Earth's atmospheric pressure. But since the main craft will be pressurized with a normal nitrogen/oxygen mixture at almost atmospheric pressure, astronauts will need to prebreathe pure oxygen (see main text). On the Space Shuttle 2½ hours of prebreathing are required; for the Space Station NASA is working on a 'hardsuit' that can contain higher pressures, in order to avoid the need for prebreathing.

The removal of exhaled carbon dioxide is a problem. In present-day spacesuits the carbon dioxide is absorbed by lithium hydroxide in canisters that are afterwards discarded. A Mars mission will need a lightweight substitute that can be regenerated after each EVA.

Before donning the spacesuit, the astronaut will pull on an undergarment that has sewn-in tubes through which cooling water circulates from the backpack. The warmed water could be cooled again by passing it through radiators on the back, or by passing it over solid carbon dioxide (dry ice) made from the Martian air.

A small chest unit will carry controls and instruments indicating levels of supplies and carbon dioxide. Inside the spacesuit helmet protruding tubes supply water and orange juice, which the astronaut will consume at an average rate of 8 ounces (250 grams) per hour. Snack food in the form of bars can make good the 750 calories consumed in an eight-hour EVA. A permanently

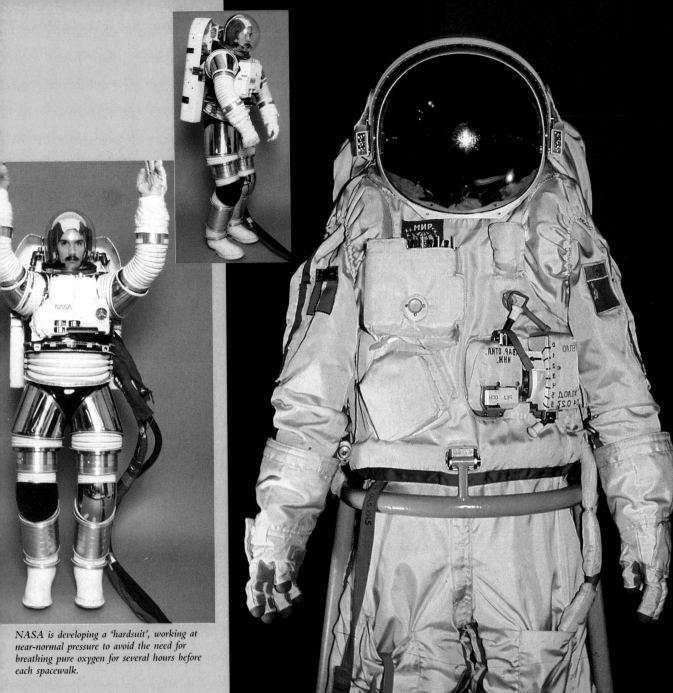

NASA is developing a 'hardsuit', working at near-normal pressure to avoid the need for breathing pure oxygen for several hours before each spacewalk.

attached tube attached to a bag carries away the 1¾ pints (1 litre) of urine produced over the same period. An absorbent undergarment deals with any bowel movement that cannot be deferred.

Conventional suits are very constricting. The Apollo Moonsuits, though designed to be worn for up to 115 hours in an emergency, were always shed with relief, even after a few hours of wear. In the longer term the 'skinsuit' may be promising for Mars bases. It would supply oxygen for breathing, but there would not be a pressurized internal atmosphere around the body. The suit would be a stretchy 'skin', the tension of which would

provide the pressure that the body requires. The essential requirement is that the pressure it exerts should be precisely equal all over the body. Preliminary trials have been encouraging but there have been problems with making the pressure even, and this has led to pooling of the blood.

There would have to be further outer layers for heat retention and protection against micrometeoroids. But such a suit would be flexible and easy to maintain, and a puncture would not be fatal in a few minutes, as it would with a pressure suit.

The suit now used by cosmonauts for spacewalks from the Soviet Mir space station would be adapted for a Mars mission.

Present-day spacesuit gloves are a particular complaint of astronauts: they are tiring on the hands and give a poor sense of touch. One Apollo astronaut even ended up with bleeding fingertips. Early Mars expeditions might adopt a suit that was conventional except that it would have skinsuit gloves. After experience has been gained with such a hybrid suit, the full skinsuit could be phased in.

PART FOUR
MYSTERIOUS PLANET

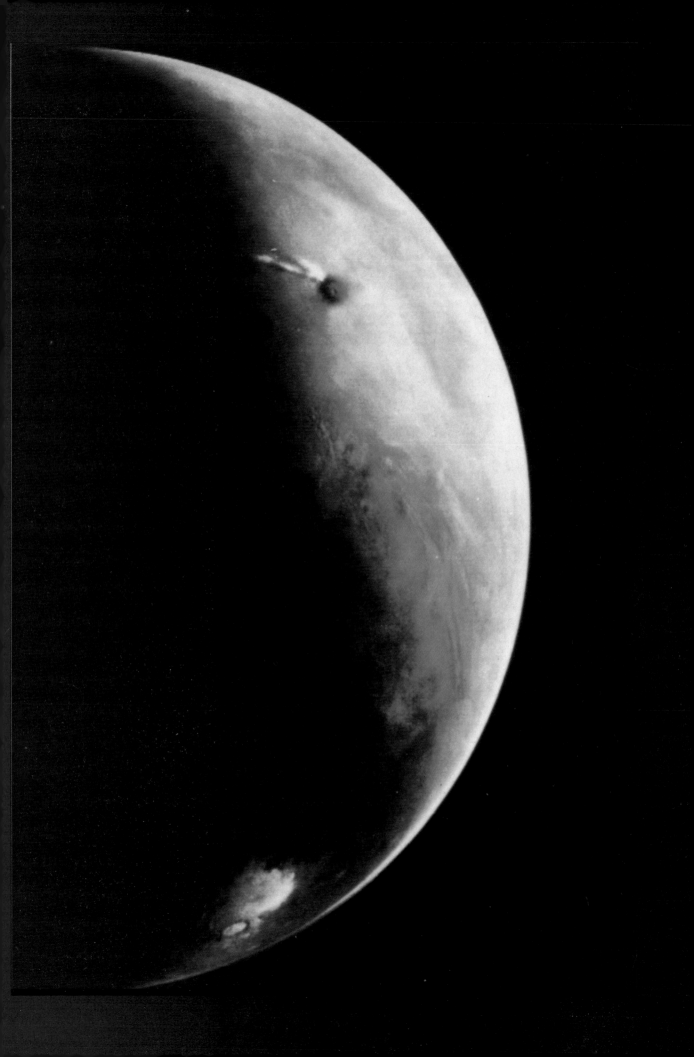

Even though it is the most Earth-like of the planets, Mars will prove far from hospitable to explorers from the Earth. For the most part it is freezing cold, with temperatures falling as low as −200°F (−130°C) at the poles. Its rarefied atmosphere consists mainly of carbon dioxide with virtually no oxygen. With a ground-level pressure less than one hundredth of the Earth's, the Martian 'air' can support thin clouds of ice crystals and dust. But it cannot trap much of the Sun's warmth, nor can it block out the Sun's deadly ultraviolet radiation. And though there is water, there is none in liquid form.

These hostile features of Mars are tempered by some similarities with the Earth. The Martian day lasts 24 hours 37 minutes, only slightly longer than ours. The planet experiences seasons, since it is tilted on its axis by 24°, roughly the same angle as the Earth. Like the Earth, Mars has ice caps at its poles. The Martian year, however, is almost twice as long as ours—687 Earth days.

Mars is much smaller than the Earth. Its diameter is 4219 miles (6790 kilometres) compared with the 7926 miles (12,756 kilometres) of our planet. Its mass—the quantity of material that makes up the planet—is about one ninth that of Earth's. The gravitational pull on the Martian surface is 40 per cent of that on Earth, which means that an explorer will experience himself as having only two-fifths of his Earth weight.

The highly elliptical orbit of Mars carries it as far as 155 million miles (249 million kilometres) from the Sun, and as close as 128 million miles (207 million kilometres). The planet moves with varying speed, being slowest when it is farthest from the Sun. This variation has a marked effect on the Martian climate. During the southern summer Mars is closest to the Sun and travelling at its fastest: the season is therefore hotter and shorter than the northern summer.

Planet of dust storms

At the edge of the southern ice cap the extreme temperature contrasts during summer cause strong winds to blow, at up to 250 miles per hour (400 kilometres per hour). Orbiters have seen features formed by strong winds, such as streamlined outcrops of rock and long wind streaks on the lee side of craters.

These winds create fierce dust storms, which can blot out the entire surface of the planet. Just such a storm was raging in November 1971 when the Mariner 9 spacecraft arrived. Mariner was destined to take the most spectacular pictures of Mars seen until then, but for many weeks its cameras showed the planet only as a featureless disc. Only when the dust settled did surface features begin to appear.

The volcanoes

At first only four dark spots, lying near the equator, could be seen above the murky layers of dust. These proved to be volcanoes, the largest of which was subsequently named Olympus Mons. It is the largest known volcano in the solar system, 17 miles (27 kilometres) high.

In general shape Olympus Mons resembles a terrestrial 'shield' volcano, broad and with gently sloping sides. But for comparison, Mauna Loa in Hawaii, the largest shield volcano on Earth, is only five miles (eight kilometres) in height and 75 miles (120 kilometres) across.

The three other volcanoes first seen by Mariner 9 form a line running diagonally across the equator. Though they are slightly smaller than Olympus Mons, their

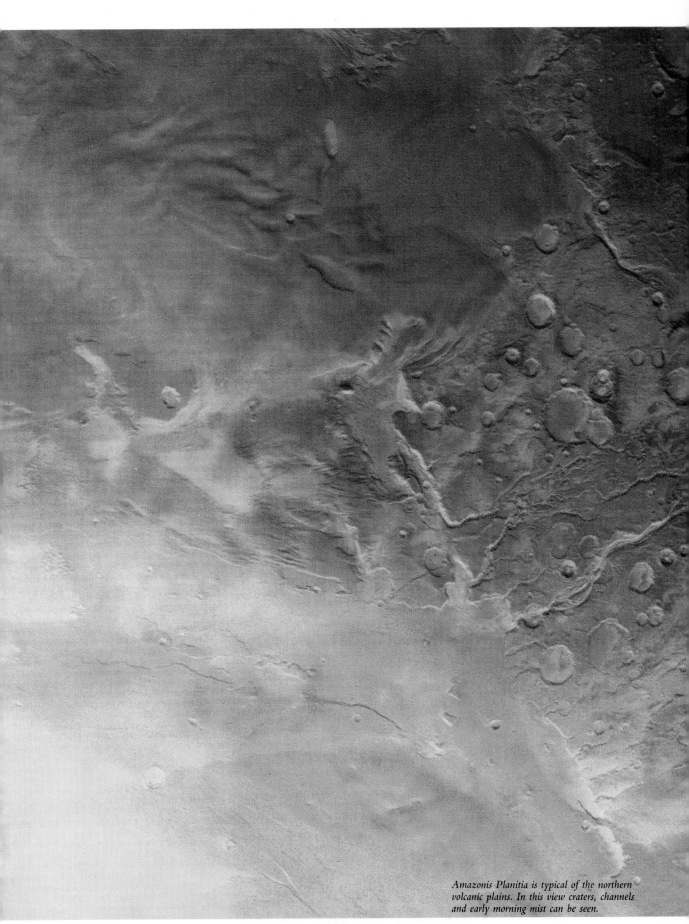

Amazonis Planitia is typical of the northern volcanic plains. In this view craters, channels and early morning mist can be seen.

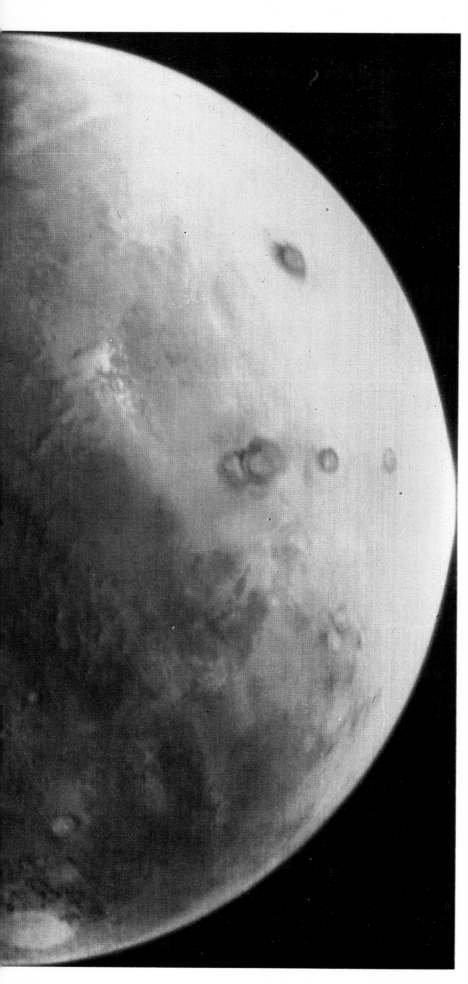

height is comparable because they lie on the Tharsis bulge. This protuberance in the Martian crust is 2200 miles (3600 kilometres) long and 5.5 miles (nine kilometres) high. Along with its volcanoes it dominates the geology of half the planet's surface.

North and south

As the rest of the planet's surface became visible to Mariner 9's cameras the northern and southern hemispheres were seen to be quite different. There is another bulge in the south, about two miles (one kilometre) high, making the planet's shape distinctly asymmetrical. And the northern hemisphere, consisting of extensive plains, is much smoother than the southern, which is heavily scarred by craters and superficially resembles the highlands of our Moon.

Generally the craters are not volcanic but meteoritic in origin and most were formed four billion years ago. The largest of the Martian craters is Hellas Basin in the southern hemisphere, 1000 miles (1600 kilometres) across.

Impact craters tell us a great deal about the development of the planets. Within the first billion years of the solar system's history—the first fifth of its life—there was a massive meteoritic bombardment of all the planets. The way in which the craters dating from that time have since been eroded and otherwise modified shows how geologically active the planet is. On Mars very few craters are 'fresh' in appearance—most have been worn down by a variety of processes over billions of years.

Equatorial canyons

Perhaps the most awe-inspiring feature on Mars is the array of huge interconnecting canyons, just south of the equator, called

The four equatorial volcanoes appear as distinct dark spots in this picture. Bright frost is visible in many craters in the southern hemisphere.

Valles Marineris in honour of Mariner 9. The system is almost 2480 miles (4000 kilometres) long, 435 miles (700 kilometres) across at its widest, and four miles (seven kilometres) deep. Transferred to the Earth, it would stretch across the United States.

The individual canyons of Valles Marineris run roughly parallel before merging into short, broken valleys, to form a type of landscape called 'chaotic terrain'. Landslides and faulting can be seen in the canyons.

Many geological processes were responsible for the creation of the canyons, though their exact origins are unclear. The dominant factor was certainly tectonism, the deformation of the planet's crust due to the flow of heat from within. The Earth's crust is divided into 'plates' whose movement powers the drifting of the continents (plate tectonics). On Mars, the crust did not evolve to form plates, but separation of the crust may have opened up the Valles Marineris network.

Clouds, fogs and frost
Unlike the Earth, 70 per cent of whose surface is covered by ocean, Mars is devoid of liquid water. The atmospheric pressure is so low that it would evaporate explosively, despite the low temperature. But water (as well as carbon dioxide) is observed as clouds, fogs and frosts. Despite the thinness of the Martian atmosphere, clouds often appear over the Martian equator at midday; the warming of the atmosphere is then greatest, causing the Martian air to rise. As it rises, it cools and can no longer hold its burden of water as vapour. The water freezes to form clouds of ice crystals. Clouds also often appear around the volcanoes in the Tharsis region, as air passing over them is forced upwards and cooled. Fogs appear at sunrise, frequently in the depths of canyons, and it seems that they consist of a mixture of water

vapour and carbon dioxide. And frost has been seen during winter by the Viking 2 lander, stationed at latitude 48°N.

The problem of water
The whole question of water on Mars presents some interesting problems. The superbly detailed pictures of Mars sent back by orbiting spacecraft give clues to a former abundance of water on the planet. The equatorial regions are

'Chaotic terrain' at the eastern end of the great equatorial canyon system, Valles Marineris.

dissected by channels reminiscent of dried-up river beds, and it is no longer doubted that they were made by running water. Many of the larger channels begin at the east of Valles Marineris and some of them extend for hundreds of miles.

But what happened to that water? There is evidence that Mars once had a denser atmosphere and a warmer, wetter climate. One reason for the change could lie in long-term alterations in the shape of the planet's orbit and the tilt of its axis. It is likely that water reserves exist in permafrost below

the Martian surface. This potential water supply will be vitally important to manned missions.

The poles
Water is also found as ice at the Martian poles. The permanent ice caps are covered in winter by a layer of frozen carbon dioxide that advances and retreats with the change of the seasons. The permanent cap at the north pole is a layer of water ice, while the smaller one at the south pole consists of a mixture of water ice and frozen carbon dioxide. Both display a spiral patterning, caused by radial winds.

The weather is significantly affected by the polar ice caps. The atmospheric pressure falls by 20 per cent whenever the carbon dioxide in the air freezes out on the winter ice cap.

The ice caps are layered because the mixture of ice and soil has been repeatedly eroded and covered by further deposits. The apparent regularity of the layers may indicate cyclic climatic changes over long epochs.

The two moons
Deimos and Phobos circle this cold dry desert world in orbits lying directly above the planet's equator. They are irregular chunks of dull grey rock, among the darkest bodies in the solar system.

Phobos, the outermost satellite, is only 12 miles (20 kilometres) across and 17 miles (28 kilometres) long, and is heavily cratered. One crater is huge in relation to the size of the satellite, being six miles (10 kilometres) across. Pronounced parallel fractures running from the crater may well have been caused by the impact that created the crater.

Deimos is smaller, only 10 miles (16 kilometres) long and six miles (10 kilometres) wide. It is not so heavily cratered as Phobos.

The origin of the two moons is uncertain. They are most likely to be captured asteroids.

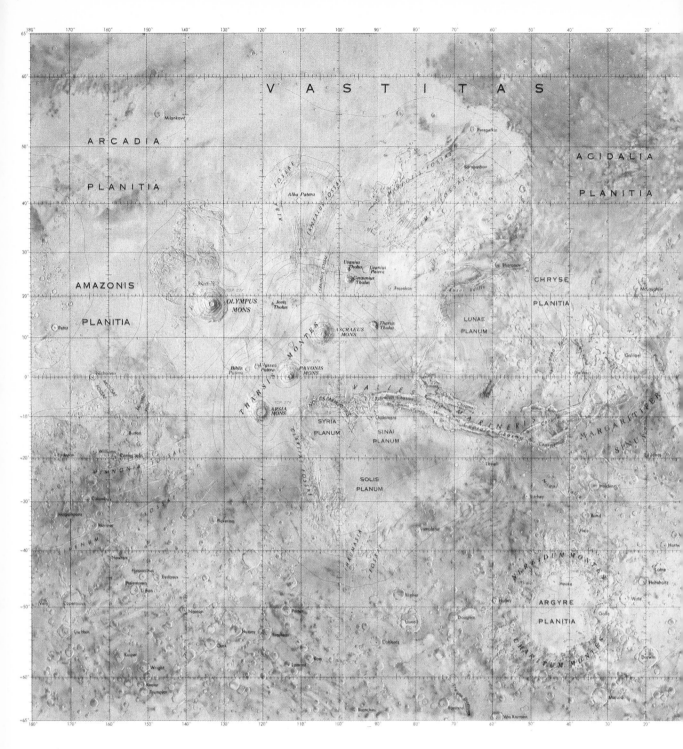

The value of planetary maps in astronomy and space exploration is manifold: by showing surface details in an appealing, uncluttered manner, they aid in determining the geology of the surface and choosing landing sites. The Mariner 9 mission first made it possible to produce accurate maps of the whole of Mars. Merely joining the Mariner images to make photomosaics was of limited value, because of the resulting patchwork-quilt effect (see page 14). This was due to variations between frames taken under differing atmospheric and lighting conditions, and at varying altitudes. The task of turning the images into maps was given to skilled cartographers at the US Geological Survey, Flagstaff, Arizona, who had developed mapping techniques for the Moon during the Apollo programme.

To produce accurate maps it is first necessary to define a grid of latitude and longitude for the particular projection to be used. The spacecraft's position is tracked continuously by radio, so the positions of the features appearing in each picture can be specified and then marked on the required map grid.

Each spacecraft image is stored in digital form. The 'best' one for a particular area is transformed by computer so that surface features are in their correct position according to the chosen projection. The computer then generates

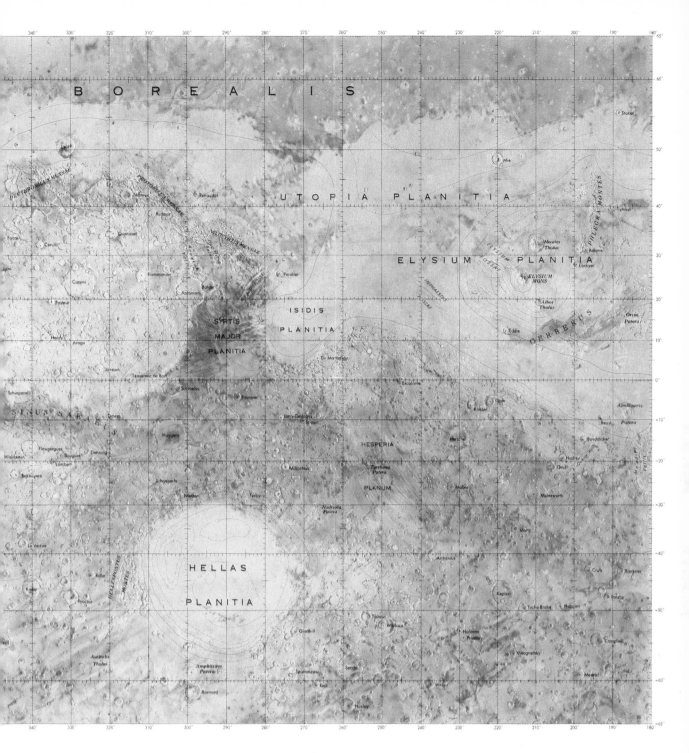

a mosaic from which the cartographers can work.

The cartographic artist's first task is to trace the main surface features from the mosaic onto clear plastic sheets, which are then placed over a white background. The artist thoroughly examines as many images as possible of the area being worked on. Then, using an artist's spraygun known as an airbrush, he or she adds further details. The airbrush is versatile enough to tint large areas or draw fine lines. An electric eraser can further modify and highlight features.

Other cartographers and scientists with a particular knowledge of the area being drawn review the work of each artist to ensure consistency. So the final maps are the result of highly skilled artists working on the photographic images, aided by complex computer programs.

The basic Mars maps are of 'shaded-relief' type, similar to 'hillshaded' maps of the Earth's surface: the shading imitates shadows cast by the features, giving an impression of the surface relief.

In addition, surface markings can be added, derived from enhancement of the spacecraft images and from measurements of the surface albedo (the proportion of the light falling on the surface that is reflected).

Preliminary maps were produced from Mariner 9 images at a scale of 1: 25 million but the data

in them were at rather poor resolution. The TV pictures sent back by the Viking orbiters enabled the US Geological Survey to produce more detailed and accurate maps of Mars. Global maps were made with a scale of 1 : 5 million—that is, one inch to 79 miles (one centimetre to 50 kilometres).

Data from Earth-based radar and the spacecraft's motion have allowed contours to be added to produce 'topographic' maps like the maps of the equatorial regions shown on page 108. Wherever possible topographic maps have been made, but in many regions the Viking pictures revealed changes in the surface markings between 1972 and the late 1970s. These made it difficult to produce definitive topographic maps. Thus the polar maps reproduced here are shaded-relief rather than topographic maps.

The US Geological Survey has produced more detailed maps of some regions, down to a scale of 1 : 250,000, each covering an area roughly 120 miles (200 kilometres) square. Future missions such as NASA's Mars Observer and the Soviet orbiters will allow even more surface detail to be mapped. Such maps will form the basis for manned missions.

Naming features

The great majority of the features now known on Mars were discovered by the Mariner 9 and

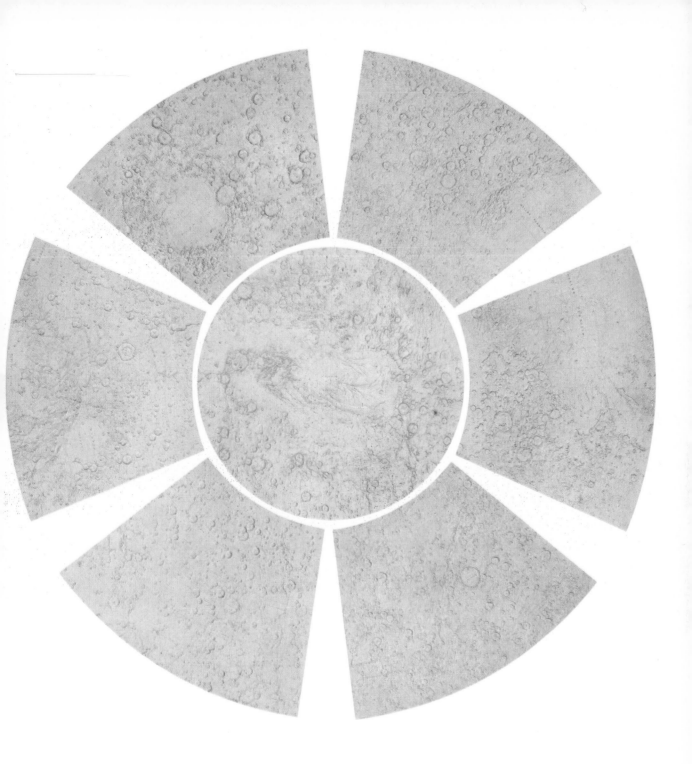

Viking missions. Wherever possible, the older names of features discovered by observers on Earth have been retained: Syrtis Major is one example. But many of the older names do not correspond to any surface feature (for example, Aethiopia). Some of the names given by Earthbound observers have been modified: Nix Olympica ('The Snows of Olympus') was given its name because it appeared as a bright spot through terrestrial telescopes. It was renamed Olympus Mons when its true nature was realized.

The nomenclature of Mars is a curious mixture. Most features are named after towns, rivers and scientists (many of them involved with Mars studies). The names of major valleys are the words for Mars in many languages (thus Mangala is Sanskrit, Kasei is Japanese). All have been approved by the International Astronomical Union, the body responsible for planetary nomenclature.

Geological features on Mars

Chasma—canyon.
Dorsum—ridge.
Fossa—ditch.
Labyrinthus—valley network.
Mensa—mesa.
Mons—mountain.
Montes—mountain range.
Patera—irregular crater.
Planitia—plain.
Planum—large plateau.
Tholus—small domed hill.
Vallis—valley or *Valles*—valleys.
Vastitas—extensive plain.

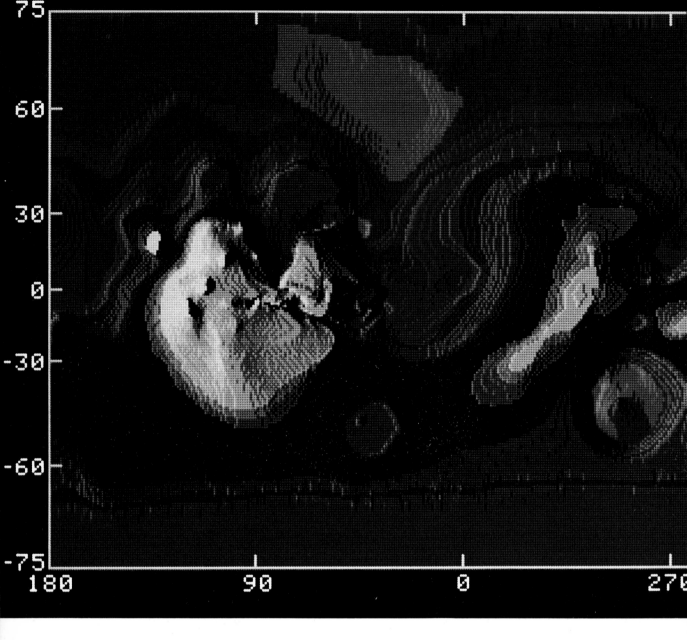

Gazetteer

	Deg. Lat.	Deg. Lon.(w)
Chasma		
Australe	80–88 S	270
Boreale	85 N	65–30
Candor	04–06 S	78–73
Capri	14–03 S	52–32
Coprates	11–14 S	68–54
Eos	16–17 S	51–32
Ganges	08 S	52–48
Hebes	01 N–01 S	81–73
Ius	07 S	98–80
Juventæ	04 S	61
Melas	08–12 S	78–70
Ophir	03–09 S	77–64
Tithonium	04 S	90–80
Dorsum		
Argyre	61–65 S	70
Fossa		
Alba	38–49 N	117–109
Cerauniæ	25 N	107
Claritas	19–32 S	108–105
Elysium	28–26 N	225–219
Hephæstus	22–18 N	240–233
Mareotis	41–48 N	85–69
Medusæ	08 S	162
Memnonia	22–15 S	158–140

	Deg. Lat.	Deg. Lon.(w)
Nili	20–26 N	284–279
Sirenum	36–27 S	163–138
Tantalus	34–47 N	105–99
Tempe	35–46 N	80–62
Thaumasia	36–40 S	100–80
Labyrinthus		
Noctis	05–08 S	110–92
Mensa		
Deuteronilus	42–45 N	346–340
Nilosyrtis	32	290
Protonilus	49–38 N	325–303
Mons		
Arsia	09 S	121
Ascræus	12 N	104
Elysium	25 N	213
Olympus	18 N	133
Pavonis	01 N	113
Montes		
Charitum	57 S	50–32
Hellesponti	45–48 S	315
Nereidum	48–38 S	57–43
Phlegra	31–46 N	195
Tharsis	12–16 N	125–101

	Deg. Lat.	Deg. Lon.(w)
Planitia		
Acidalia	48 N	30
Arcadia	48 N	155
Amazonis	13 N	160
Argyre	49 S	43
Chryse	17 N	45
Elysium	15 N	210
Hellas	45 S	290
Isidis	15 N	270
Syrtis Major	15 N	290
Utopia	35 N	235
Planum		
Auroræ	10–11 S	52–48
Hesperia	10–35 S	258–242
Lunae	05–20 N	70–60
Ophir	09–12 S	61–55
Sinai	10–20 S	90–70
Solis	20–30 S	98–88
Syria	10–18 S	105–100
Patera		
Alba	40 N	110
Amphitrites	59 S	299
Apollinaris	08 S	186
Biblis	02 N	124
Hadriaca	31 S	267

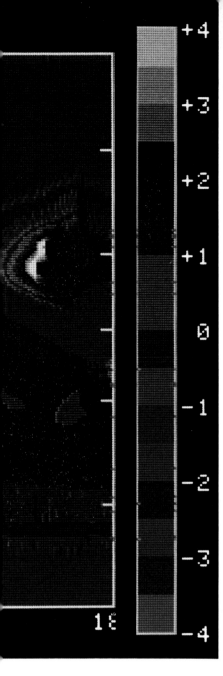

+4
+3
+2
+1
0
−1
−2
−3
−4

18

	Deg. Lat.	Deg. Lon.(W)
Orcus	14 N	181
Tyrrhena	22 S	253
Ulysses	03 N	121
Uranius	26 N	93
Tholus		
Albor	19 N	210
Australis	57 S	323
Ceraunius	24 N	97
Hecates	32 N	210
Iaxartes	72 N	15
Jovis	18 N	117
Kison	73 N	358
Ortygia	70 N	8
Tharsis	14 N	91
Uranius	26 N	98
Vallis		
Al Qahira	23–15 S	202–194
Ares	02–10 N	23–14
Auqakuh	28 N	298
Huo Hsing	32–28 N	295–292
Kasei	21 N	70–56
Ma'adim	27–20 S	183
Mangala	10–4 S	151
Marineris	05–15 S	96–45
Nirgal	32–27 S	44–36
Shalbatana	01–15 N	45
Simud	00–14 N	40–37
Tiu	10–18 N	32

	Deg. Lat.	Deg. Lon.(W)
Vastitas		
Borealis	55–67 N	–
Craters		
Adams	31 N	197
Agassiz	70 S	83
Airy	0·5 S	0
Antoniadi	22 N	299
Arago	10 N	330
Arrhenius	40 S	237
Bakhuysen	23 S	344
Baldet	23 N	295
Barabashov	47 N	69
Barnard	61 S	298
Becquerel	22 N	8
Beer	15 S	8
Bianchini	64 S	97
Bjerknes	43 S	189
Boeddicker	15 S	197
Bond	33 S	36
Bouguer	19 S	333
Brashear	54 S	120
Briault	10 S	270
Burroughs	72 S	243
Burton	14 S	156
Campbell	54 S	195
Cassini	24 N	328
Cerulli	32 N	338
Chamberlain	66 S	124
Charlier	69 S	169
Clark	56 S	134
Coblentz	55 S	91
Columbus	29 S	166
Comas Solá	20 S	158
Copernicus	50 S	169
Crommelin	5 N	10
Cruls	43 S	197
Curie	29 N	5
Daly	66 S	22
Dana	73 S	32
Darwin	57 S	20
Dawes	9 S	322
Denning	18 S	326
Douglass	52 S	70
Du Martheray	6 S	266
Du Toit	72 S	46
Eddie	12 N	218
Ejriksson	19 S	174
Escalante	0	245
Eudoxus	44 S	147
Fesenkov	22 N	87
Flammarion	26 N	312
Flaugergues	17 S	341
Focas	34 N	347
Fontana	64 S	73
Fournier	4 S	287
Gale	6 S	222
Galilei	6 N	27
Galle	51 S	31
Gilbert	68 S	274
Gill	16 S	354
Gledhill	53 S	273
Graff	21 S	206
Green	52 S	8
Hadley	19 S	203
Haldane	53 S	231
Hale	36 S	36
Halley	49 S	59
Hartwig	39 S	16
Heaviside	71 S	95
Helmholtz	46 S	21
Henry	11 N	336
Herschel	14 S	230
Hipparchus	44 S	151
Holden	26 S	34
Holmes	75 S	292
Hooke	45 S	44
Huggins	49 S	204
Hussey	54 S	127
Hutton	72 S	255
Huxley	63 S	259
Huygens	14 S	304
Janssen	3 N	322
Jarry-Desloges	9 S	276
Jeans	70 S	206
Joly	75 S	42
Jones	19 S	20
Kaiser	46 S	340
von Kármán	64 S	59
Keeler	61 S	152
Kepler	47 S	219
Knobel	6 S	226
Korolev	73 N	196
Kuiper	57 S	157
Kunowsky	57 N	9
Lambert	20 S	335
Lamont	59 S	114

	Deg. Lat.	Deg. Lon.(W)
Lampland	36 S	79
Lassell	21 S	63
Lau	74 S	107
Le Verrier	38 S	343
Liais	75 S	253
Li Fan	47 S	153
Liu Hsin	53 S	172
Lockyer	28 N	199
Lomonosov	65 N	8
Lowell	52 S	81
Lyell	70 S	15
Lyot	50 N	331
McLaughlin	22 N	22
Mädler	11 S	357
Magelhaens	32 S	174
Maggini	28 N	350
Main	77 S	310
Maraldi	62 S	32
Mariner	35 S	164
Marth	13 N	3
Martz	34 S	217
Maunder	50 S	358
Mendel	59 S	199
Mie	48 N	220
Milankovič	55 N	147
Millochau	21 S	275
Mitchel	68 S	284
Molesworth	28 S	211
Moreux	42 N	315
Müller	26 S	232
Nansen	50 S	141
Newcomb	24 S	358
Newton	40 S	158
Nicholson	0	166
Niesten	28 S	302
Oudemans	10 S	92
Pasteur	19 N	335
Perepelkin	52 N	65
Peridier	26 N	276
Pettit	12 N	174
Phillips	67 S	45
Pickering	34 S	133
Playfair	78 S	125
Porter	50 S	114
Priestley	54 S	228
Proctor	48 S	330
Ptolemæus	46 S	158
Quénisset	34 N	319
Rabe	44 S	325
Radau	17 N	5
Rayleigh	76 S	240
Redi	61 S	267
Renaudot	42 N	297
Reuyl	10 S	193
Reynolds	74 S	160
Richardson	73 S	181
Ritchey	29 S	51
Ross	58 S	108
Rossby	48 S	192
Rudaux	38 N	309
Russell	55 S	348
Rutherford	19 N	11
Schaeberle	24 S	310
Schiaparelli	3 S	343
Schmidt	72 S	79
Schroeter	2 S	304
Secchi	58 S	258
Sharonov	27 N	59
Sklodowska	34 N	3
Slipher	48 S	84
Smith	66 S	103
South	77 S	339
Spallanzani	58 S	273
Steno	68 S	115
Stokes	56 N	189
Stoney	69 S	134
Suess	67 S	179
Teisserenc de Bort	1 N	315
Terby	28 S	286
Tikhov	51 S	254
Trouvelot	16 N	13
Trumpler	62 S	151
Tycho Brahe	50 S	214
Tyndall	40 N	190
Very	50 S	177
da Vinci	2 N	39
Vinogradsky	56 S	217
Vishniac	77 S	276
Vogel	37 S	13
Wallace	53 S	249
Wegener	65 S	4
Weinbaum	66 S	245
Wells	60 S	238
Williams	18 S	164
Wirtz	49 S	26
Wiuslencius	18 S	349
Wright	59 S	151

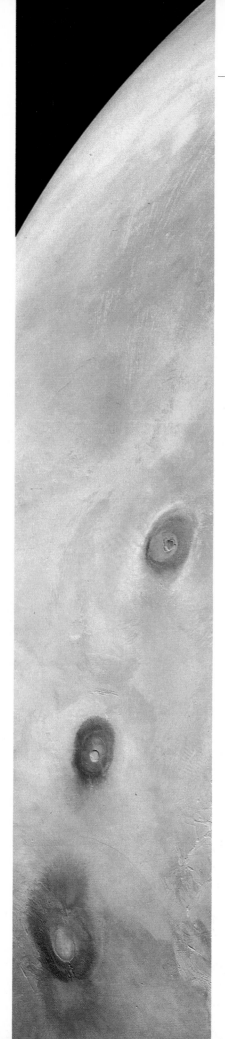

VOLCANOES

Volcanoes are the most staggering geological features on Mars. The most spectacular are to be found in the equatorial region of Tharsis, but there are many smaller volcanoes right across the planet. Volcanism has taken place throughout much of the planet's history, the most important events being the formation of the vast northern lava plains and the development of the Tharsis volcanoes. As we have seen (page 100), the latter are shield volcanoes, and their similarity in appearance suggests they had similar eruptive histories. Their overall shape—with shallow slopes and calderas on their summits—suggest eruptions of fluid lava, with very little ash content. As with terrestrial volcanoes, the chemical composition of the lava and the rate at which it erupted affected the volcano's eventual structure.

Shield volcanoes

Calderas or 'volcanic craters' are caused by the collapse of the summit after lava has flowed. Before an eruption, pressure builds up within the volcano, making it 'inflate'. Once there has been a flow of lava, the pressure is relieved, and the magma underneath the summit then withdraws. The summit 'deflates' and a depression is formed—a caldera. If this happens regularly, the caldera assumes a complex 'nested' appearance—a feature of many Martian shield volcanoes. The cycles of inflation and deflation within, for example, the Tharsis calderas are similar to those that occurred in Hawaiian volcanoes.

But the Tharsis volcanoes are between 10 and 100 times larger than the Hawaiian volcanoes, their nearest terrestrial equivalents. Why are they so big?

One reason may be that eruption rates were probably higher at the time of the volcanoes' for-

mation. But more important is the fact that Mars lacks plate tectonics. The Earth's crust is made up of a number of plates, which are in constant movement. Volcanoes on the Earth are most commonly formed where plates are colliding or at hot spots where plates are separating. The great Hawaiian shield volcanoes grew above hot spots under the Pacific plate. Each could grow only to a certain size before it was carried away by the movement of the plate, and became extinct. Another volcano formed over the hot spot, repeating the process. In this way a chain of extinct volcanoes was formed, stretching from Hawaii to Midway Island. On Mars the crust is much thicker, and it remains static. A volcano remains above a hot spot for many millions of years, swelling into the gigantic structures we see today.

What sort of lava formed the vast volcanic plains on Mars? Geologists believe it was 'basaltic' rock, a type common to many of the inner planets. Basaltic rocks are relatively dense, and are rich in both magnesium and iron. Basaltic lavas erupt at temperatures around 1800°F (1000°C) and so are fairly fluid. They would have been able to flow across the surface for many hundreds of miles before cooling.

Olympus Mons

The largest volcano in the solar system is named, appropriately, after the abode of the Greek gods. It is truly enormous, towering some 17 miles (27 kilometres) above the local terrain, and spanning some 370 miles (600 kilometres) at its base. Mount Everest, to provide a terrestrial comparison, rises only 5.5 miles (9 kilometres) above sea level. The caldera on top of Olympus is about 50 miles (80 kilometres) across, and within it are several nested volcanic craters, the deepest of which is 26 miles (42

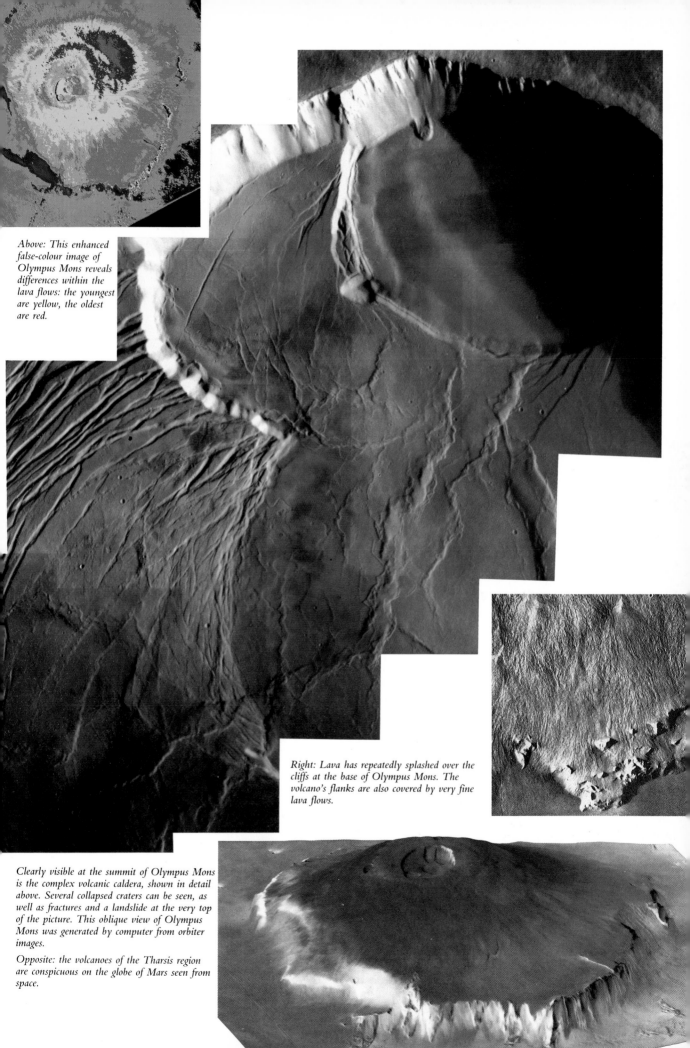

Above: This enhanced false-colour image of Olympus Mons reveals differences within the lava flows: the youngest are yellow, the oldest are red.

Right: Lava has repeatedly splashed over the cliffs at the base of Olympus Mons. The volcano's flanks are also covered by very fine lava flows.

Clearly visible at the summit of Olympus Mons is the complex volcanic caldera, shown in detail above. Several collapsed craters can be seen, as well as fractures and a landslide at the very top of the picture. This oblique view of Olympus Mons was generated by computer from orbiter images.

Opposite: the volcanoes of the Tharsis region are conspicuous on the globe of Mars seen from space.

kilometres) across. Many of these characteristically nested craters have collapsed and coalesced, and show wrinkled edges on their floors from lava flows. The caldera walls are also fluted, indicating much landslide activity.

The lava itself has spilled out onto the surface in fingerlike flows hundreds of miles long. The size of the Olympus Mons caldera testifies to the vastness of the magma reservoir beneath it; the long lava flows show that the rates of eruption were very high. These two factors were paramount in the impressive growth of Olympus Mons.

On its flanks the surface texture is very fine, produced by the collapse of small channels and lava tubes. Lava flows overlap and bury each other, so it is not clear whether any eruptive vents or fissures opened up, as is common in some terrestrial volcanoes. Lava tubes are common in Hawaii and Iceland, but are rarely more than 35 feet (10 metres) in width. On Mars similar features often exceed 650 feet (200 metres) across. On Mars large lava flows occupy four

The complex structure of the caldera on Arsia Mons, southernmost volcano in the Tharsis region, has resulted from the repeated formation and collapse of craters.

to 10 times the volume of equivalent flows on the Earth.

Skirting the base of Olympus Mons is a scarp, or cliff, with a depth of 2.5–3.5 miles (4–6 kilometres). This feature would dominate the horizon seen by explorers on the surface. But the volcano as a whole would not look spectacular, since its flanks slope at only about 4°, a typical value for shield volcanoes. Lava has repeatedly splashed over the cliff's edge, giving it a smooth appearance.

The volcano as a whole is surrounded by an incomplete ring of coarsely ridged, hummocky surface rock. This 'aureole', as it is known, extends for 180–420 miles (300–700 kilometres) beyond the scarp, and is something of a mystery to geologists. Some see it as the remains of an even bigger, more ancient shield volcano; others think it may have been caused by great landslides. Whatever the truth may be, the region will provide a fascinating survey site for later explorers on Mars.

The Tharsis volcanoes

To the south-east of Olympus Mons are three large volcanoes in a line: from north-east to south-west they are Ascraeus Mons,

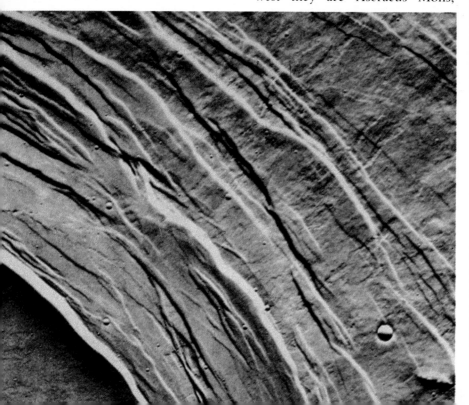

Pavonis Mons and Arsia Mons. They are roughly six miles (10 kilometres) high, and 210–280 miles (350–450 kilometres) wide at their bases. Each is about 420 miles (700 kilometres) from its neighbour. Because they lie on top of the Tharsis bulge, which is itself some 5.5 miles (9 kilometres) above the surrounding terrain, the Tharsis volcanoes are at the same height as Olympus Mons. This may be no coincidence: it might be the maximum height they could build up to before the crust beneath became unstable.

Above: This view across Tharsis shows Pavonis Mons (centre) and Arsia Mons (bottom). Noctis Labyrinthus, an intricate system of canyons, is seen at top right.

Right: Irregular lava fissures emanate from the flanks of Arsia Mons.

Though the three volcanoes have roughly the same morphology—suggesting they underwent the same growth cycles—the calderas at their summits have different appearances. The caldera on Arsia Mons is 68 miles (110 kilometres) in diameter, and around

1.2 miles (two kilometres) deep.
(By comparison the caldera of
Mauna Loa in Hawaii is a mere 1.7
miles [2.7 kilometres] across and
only 650 feet [200 metres] deep.)
To the south-west flank of Arsia
are irregular depressions, caused
by repeated lava eruptions from
vents in the side of the volcano.
Later in its development, most of
the lava emanated from such
vents.

Pavonis Mons has a much simpler, circular caldera, 2.5–3 miles
(4–5 kilometres) deep. Its walls are
terraced, suggesting the caldera as
a whole collapsed repeatedly. The
caldera on Ascraeus Mons is much
more complex: shadow measurements show that it is about 2.5
miles (four kilometres) deep.
There are eight 'nested' craters
inside the main caldera, ranging
from 4.5 miles (7 kilometres) to 25
miles (40 kilometres) in diameter.
In many of them it is clear that
there were lakes of liquid lava at
one time.

At the north-east margins of the
Tharsis bulge a number of small
domelike volcanoes are found.
They are typically 60 miles (100
kilometres) across, and were
formed when the bulge lifted up
the crust. Hardly surprisingly, the
crust fractured and a spectacular
series of faults resulted, known as
Ceraunius Fossae.

Elysium

Also in the northern hemisphere is
a region known as Elysium,
another distinct bulge in the
Martian crust. Though smaller
than Tharsis it stands 2.5 miles
(four kilometres) above the surface
and is 1200 miles (2000 kilometres)
across. The volcanoes found here
are also much smaller than their
counterparts on Tharsis: they have
smoother sides and steeper flanks
than the Tharsis shields. This suggests that a different type of
eruption took place, probably with
a greater ash content.

Elysium Mons is the tallest of
these volcanoes, standing 5.5 miles
(nine kilometres) high. Around its

flanks are curious, parallel-sided
troughs that merge with each
other. Their appearance suggests
they have been carved by water—
perhaps produced by the melting
of subsurface ice. But an
alternative explanation is that they
may have been carved out by
extremely fluid lava.

The other two Elysium volcanoes are Hecates Tholus and
Albor Tholus—'tholus' denotes a
steep-sided structure. Very little is
known about Albor Tholus.
Hecates Tholus is 110 miles (180
kilometres) across and about 3.5
miles (six kilometres) high. It is
dome-shaped, with a flat top and a
complex caldera.

114

The patera

Another type of Martian volcano is called 'patera', and has a very low, shallow form. Perhaps the most remarkable is Alba Patera, due north of the Tharsis bulge. Its base is 1000 miles (1600 kilometres) wide, making it the most extensive Martian volcano. However, it stands only 3.5 miles (six kilometres) high, so that its sides slope at only half a degree. From the surface this sprawling structure would not appear very impressive.

At the centre of Alba Patera is the ubiquitous caldera, surrounded by many concentric fractures, which encircle the volcano as a whole. Many well preserved lava flows extend across Alba's flanks, some of them hundreds of miles long. These have given geologists the chance to compare lava flows on Earth with those on Mars. An idea of the sheer volume and ferocity of volcanic eruptions on Mars may be gleaned from a comparison of Alba Patera and Hawaiian volcanoes. In a typical Hawaiian eruption, it has been estimated, lava flows at the rate of 1750 cubic feet (50 cubic metres) per second. The rate of lava flow from Alba would have been 20 to 1000 times greater.

The oldest volcanoes on Mars are found in the southern hemisphere, in the vicinity of the Hellas basin. They resulted from the turmoil in the crust when Hellas was formed by the impact of a large body. These volcanoes are also of the patera type, and have been heavily cratered and eroded. One example is Tyrrhena Patera, which has a caldera seven miles (12 kilometres) wide, surrounded by deposits extending 180 miles (300 kilometres) from the centre. These are dissected by numerous channels, suggesting that this part of the surface is made up of material more easily eroded than that thrown out by the younger volcanoes. Perhaps the material that erupted here was more ashlike, for this is more easily eroded than rock formed from more fluid lava.

The history of Martian volcanism

Planetary scientists attempt to date Martian volcanoes by counting the numbers of impact craters on them (see page 116). One problem in doing this is that volcanic and impact craters are hard to tell apart: near the summit of Olympus Mons is a crater that was obviously caused by an impact, but on other volcanoes there is little to distinguish such craters from those that have been caused by eruptions. But rough crater counts show that the patera-type volcanoes in the southern hemisphere are the oldest, the shield volcanoes in Tharsis the youngest. Olympus Mons seems to be the youngest of all, but even so has been dormant for many millions of years. Crater counts around other volcanoes suggest that volcanic activity gradually decreased with time.

The main events in Martian volcanic history were the formation of the vast lava plains and the growth of the shield volcanoes in Tharsis. The former took place earlier than 2.5 billion years ago, while activity around Tharsis continued up to a billion years ago. It is difficult to know how much material erupted from the volcanoes and at what date; so it is no suprise that volcanoes figure prominently among possible landing sites. When the first samples of volcanic material are analysed, it will give scientists a chance to unravel the Martian past, and to glimpse conditions inside the planet.

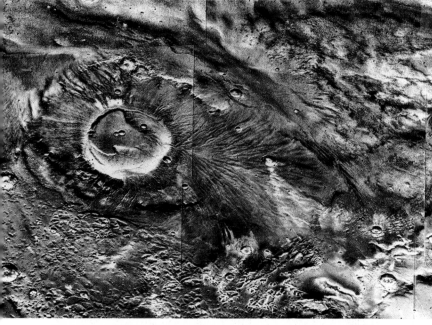

The oldest volcanoes on Mars are found in the southern hemisphere. Top: Apollinaris Patera has a caldera 60 miles (100km) wide. Lava flows stretch from the volcano for 120 miles (200km) towards the east. Above: Tyrrhena Patera and Hadriaca Patera (right and left respectively) are the most ancient of Martian volcanoes, so heavily eroded that they are hardly distinguishable against the surrounding terrain of the Hellas basin.

CRATERS AND PLAINS

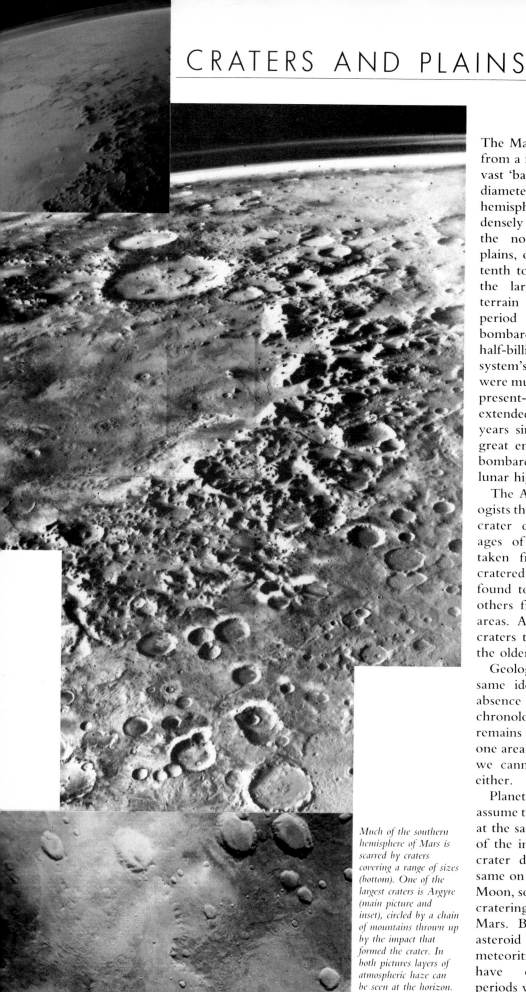

The Martian craters range in size from a few tens of yards across to vast 'basins', hundreds of miles in diameter. Most of the southern hemisphere consists of old terrain, densely scarred by craters, while in the north there are extensive plains, on which there are only a tenth to a hundredth as many of the larger craters. The oldest terrain probably records the period of intensive meteoritic bombardment during the first half-billion years of the solar system's life, when cratering rates were much higher than today. The present-day rate, even when extended over the four billion years since, would not be nearly great enough to create a heavily bombarded terrain like that of the lunar highlands.

The Apollo missions gave geologists the chance to correlate lunar crater counts directly with the ages of surface samples. Rocks taken from the Moon's lightly cratered volcanic plains were found to be much younger than others from the heavily cratered areas. And in general, the more craters there are in a given area, the older the surface is.

Geologists attempt to apply the same idea to Mars. But in the absence of Martian samples, any chronology of the Martian surface remains relative: we can say that one area is older than another, but we cannot assign actual ages to either.

Planetary geologists generally assume that cratering has occurred at the same rates on all the bodies of the inner solar system. In fact crater densities are roughly the same on parts of Mercury and the Moon, so it is highly likely that the cratering rates were the same for Mars. But being nearer to the asteroid belt, a potential source of meteoritic material, Mars may have experienced occasional periods when cratering rates were

Much of the southern hemisphere of Mars is scarred by craters covering a range of sizes (bottom). One of the largest craters is Argyre (main picture and inset), circled by a chain of mountains thrown up by the impact that formed the crater. In both pictures layers of atmospheric haze can be seen at the horizon.

higher. There is no way of being sure until samples from the Martian surface are analysed.

Another barrier to absolute dating of Martian features is the fact that many craters have been eroded or buried by wind and water erosion—processes not occurring on the Moon or Mercury.

Impact craters

Martian craters are strikingly different from their counterparts on the Moon and Mercury. The general mechanism of crater formation is understood reasonably well. Meteorites or asteroids slamming into a planetary surface can have velocities of more than six miles per second (10 kilometres per second). A body only 100 feet (30 metres) across can make a crater over 1000 yards (one kilometre) in diameter. Theoretical studies and experiments with hypervelocity guns and high-energy explosions have yielded some understanding of what happens in these extreme situations. They have shown how the impacting body is destroyed, vaporized in a few moments.

The crater Yuty, 11 miles (18km) in diameter, has a central peak and is surrounded by complex lobes of ejected material. This ejecta was very fluid, owing to the melting of subsurface ice.

Some of the surface material is also vaporized and some is melted, but these processes are of minor importance: the excavation of the crater is mostly accomplished by shock waves. As the impacting body hits the ground, its kinetic energy—energy of motion—is rapidly transferred into the surface material, compressing it and generating a high-pressure shock wave in the form of a rapidly expanding hemispherical shell. A decompression wave follows in its wake, which throws material upwards and out of the crater, spreading it across the surface to form an 'ejecta blanket'. This can cover vast areas, depending on the size and velocity of the impacting body. Very large meteorites can throw out huge clumps of ejecta, which create further smaller impact craters, called 'secondary craters'.

For a body of a given mass, the damage produced increases rapidly with increasing impact speed. The size and shape of the crater also depend on the nature of the surface material and the strength of the planet's gravity: in a stronger gravitational field the ejecta will travel a smaller distance. But hypervelocity gun experiments have shown that the angle at which the body strikes has little effect on the shape of the crater.

Craters smaller than a few miles across are usually bowl-shaped. Larger ones are much flatter, with terraced walls and a pronounced central peak. Basins—the largest craters, hundreds of miles across—do not have such peaks, but are surrounded by multiple concentric rings of mountains. These products of catastrophic impacts have been seen not only on Mars but on Mercury, the Moon, and Jupiter's satellite Callisto.

Unique craters

As gravity is an important factor in shaping craters, why aren't Martian craters similar to, say, those on Mercury, which has roughly the same surface gravity?

The main difference lies in the way in which the ejecta has covered the surrounding terrain. On Mars the ground rock and soil contain subsurface water ice: when the ejecta was flung out from the impact, the ice would have melted, allowing the material to become fluid. This did not happen on the Moon and Mercury, both arid bodies (Apollo rock samples show no sign of water on the lunar surface). The presence of the Martian atmosphere, which was probably denser in past epochs, would also have affected the motion of the ejecta.

The ejecta in lunar craters, unimpeded by an atmosphere, would have been thrown out 'ballistically' and would have crashed down onto the surface. On Mars, by contrast, the ejecta, made fluid by the melted permafrost, could have moved across the surface for greater distances after landing. Most Martian craters have many layers of ejected material, with distinct edges or 'ramparts'. Features seen within the 'ejecta blanket' include striations, radiating like bicycle spokes from the crater's centre, as well as concentric grooves and ridges. At first geologists thought that these had been caused by wind erosion, but scattered among these features there are secondary craters produced by the impact, and they do not show the effects of erosion.

The regions around larger craters would be difficult terrain to explore. Unmanned rovers would have difficulty in negotiating the rugged ejecta blankets: geologists would make better investigators in such areas.

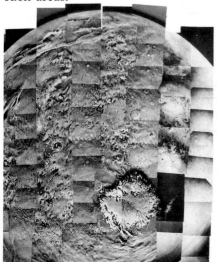

The similarity of the two Viking lander sites was surprising: they had been chosen because in orbital pictures they seemed to have differing geology. Surface rocks could result from such varied sources as lava flows, flood channels and crater ejecta. The pictures taken by the landers show craters on the horizon, suggesting that cratering is actually the most important factor and accounts for the similarity of the landscapes.

One type of erosion is missing on Mars. On the Moon continual micrometeoroid impact has worn down the surface rocks: on Mars, the surface has largely been protected by the thin atmosphere in which micrometeoroids burn up.

The basins

There are 16 impact craters with diameter greater than 150 miles (250 kilometres) on the surface of Mars. They have the appearance of vast, multi-ringed basins; some are fresh, others are old and extensively eroded.

The largest basin is called Hellas, and on maps of Mars (see pages 104–7) it dominates the southern hemisphere. It is about 1000 miles (1600 kilometres) across and is surrounded by a partially buried 'rim'. Its precise size is difficult to measure because it has been eroded in the north-east and the south-west. The rim is the first of five concentric rings of mountains, produced by the deformation of the crust when an asteroid-sized body hit the planet. Not sur-

A panorama across Chryse Planitia (above) and a mosaic of the southern hemisphere (below left).

prisingly, the effect on the surrounding terrain was enormous: two large volcanoes, Amphitrites Patera and Hadriacus Patera, seem to have been formed out of the concentric fractures. Judging by crater counts, the event that created Hellas dates from *after* the epoch of heavy bombardment.

Hellas is some three miles (five kilometres) deep, and is the lowest point on Mars. Frost often forms inside it, and at such times it can often be seen as a bright area by astronomers on Earth. It acts as a giant dustbowl: the seasonal major duststorms start in the vicinity of Hellas (see page 100). The basin's floor, known as Hellas Planitia, has been severely scoured by the strong, dust-laden winds of the southern hemisphere.

Perhaps the best-preserved of the large basins is Argyre, again in the southern hemisphere. It is about 360 miles (600 kilometres) across, and it is surrounded by a fresh-looking rim and mountain chains. Though a number of large craters have been superimposed on it, the freshness of the rim suggests that it is younger than Hellas.

Why are there so many large basins on Mars? It is impossible to determine erosion rates accurately, but it is plausible that smaller craters would disappear more rapidly than larger ones. The larger craters are more prominent because they are harder to erase.

The northern plains

There are many surprising features on the plains of the northern hemisphere. Lava flows, wind erosion and, in the past, water and ice erosion have all played a part in shaping them. Those within 30° of the equator are lava plains, resulting from the volcanoes of the Tharsis bulge (see page 100) and are fairly recent. North of 30° the landforms are much more complicated and not well understood: it is difficult to know which processes have been dominant.

Unusual craters are seen in the band between 40° and 60° north. They are termed 'pedestal' craters because they are elevated above

The landing sites of the two Viking craft, both in the northern plains. Computer enhancement highlights sand drifts seen by Viking 1 (right) and frost seen by Viking 2 (below).

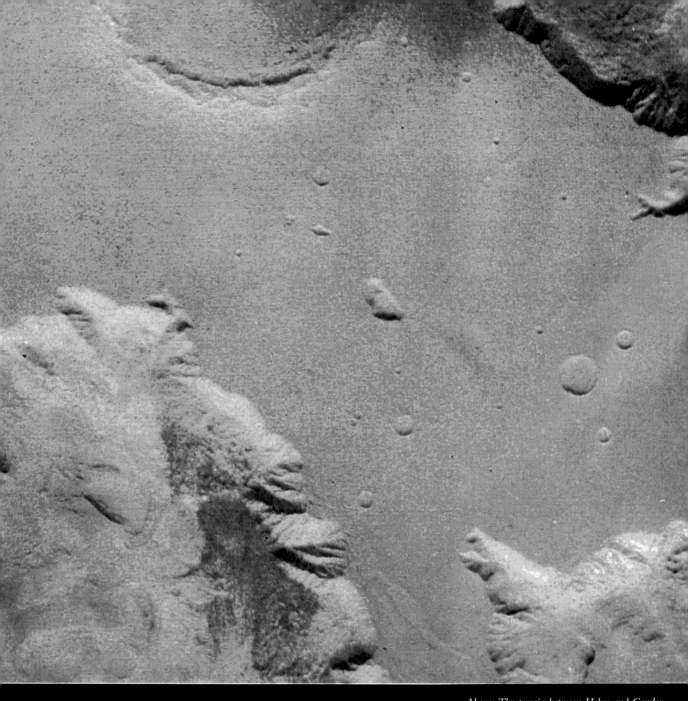

Above: The terrain between Hebes and Candor Chasma in the great equatorial canyon system. The plateau between the canyons is sparsely cratered. Colour differences can be seen in the canyon walls and floors, indicating different surface compositions.

Left: On the margins between the heavily cratered terrain and the northern plains is Nilosyrtis Mensa, which has a curiously 'fretted' appearance. This is probably due to glaciers, which carved out valleys while leaving behind ridges that are remnants of the previously existing plateau.

the surrounding land. The ground outside their walls has been worn away by wind. In other areas, repeated burying and stripping of surface material has resulted in the extraordinary sight of ejecta blankets lying below the level of the surrounding land.

In the Cydonia region strange geometrical markings resembling giant ploughed fields have been observed. Low ridges and valleys, about half a mile (800 metres) from crest to crest, wind in parallel contours to form a 'thumbprint' pattern. They are not well understood, but subsurface permafrost may have been the cause.

Streaks and splotches
There is much evidence of wind erosion on the surface of Mars. Many structures have peculiar shapes as a result of it. The strong winds are also revealed in the light and dark streaks extending from obstacles, particularly craters. Some of these streaks are a few miles wide and tens of miles in length.

Wind-tunnel experiments have shown roughly how these streaks are formed. As air flows over and around a crater, it separates into two vortices, which then converge behind the crater. The greatest erosion occurs in the areas over which these vortices flow, throwing up dust. A shadow zone is left directly behind the crater where this dust is deposited. The streaks observed may be due to erosion or deposition, depending on the wind conditions.

Irregular dark splotches are seen in some areas, particularly higher southern latitudes. They occur mostly in craters, and again seem to be caused by dust deposition and erosion during dust storms.

The directions of the streaks reveal much about the wind patterns in the lower atmosphere. Bright streaks are the most stable, while darker ones change in the course of a Martian year. In equatorial regions they tend to run

north-westerly in areas south of the equator, and north-easterly in areas north of it. These markings disclose the wind patterns that occur during the southern summer, when winds are at their strongest. Only strong winds can form such streaks, because the pressure of the rarefied atmosphere is so low.

Dunefields, caused by wind erosion (right), and atmospheric fogs and cloud (below) are commonly seen on Mars.

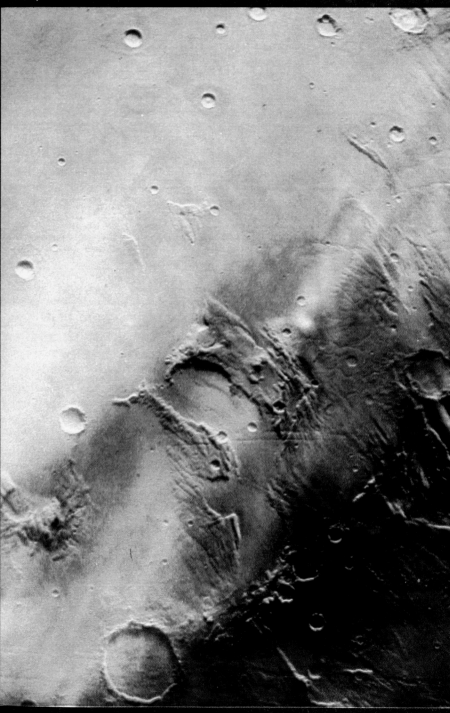

THE GREAT CANYONS

The complicated network of canyons that make up the Valles Marineris stretches for 2400 miles (4000 kilometres), and is prominent in many Viking orbiter pictures. Seen from the surface, it would stretch far over the horizon. Individual canyons within the system can be 120 miles (200 kilometres) wide and four miles (seven kilometres) or more deep. By comparison, Arizona's Grand Canyon is only 280 miles (450 kilometres) long, 1¼ miles (two kilometres) deep and at most 18 miles (30 kilometres) wide.

Complex origins

The western extremities of Valles Marineris lie on the Tharsis bulge, which can hardly be a coincidence. When the bulge was uplifted, magma may have withdrawn from around it, causing the surface to fracture and triggering earthquakes and extensive faulting. Such a tectonic origin is supported by the fact that the canyons radiate outwards from the centre of the bulge.

The Valles Marineris system does not form a well connected drainage pattern: some of the individual canyons are completely boxed in, so that water could not have flowed in or out of them. In fact, some geologists feel that the name 'canyon' is misleading: most canyons on Earth have been carved by running water.

And yet at the system's eastern edge, the canyons merge into chaotic terrain. This is a peculiarly Martian landform, which seems to have been caused by the collapse of the surface—possibly owing to the melting of subsurface permafrost—to produce a characteristic jumbled appearance. A number of the dried-up water channels emerge from this chaotic terrain (see page 103). Again, it can hardly be coincidence that the Valles Marineris system is so intimately linked to a region that seems to have been shaped by water.

The system thus presents a paradox to the geologist. It seems that complex multiple causes have been at work in its formation.

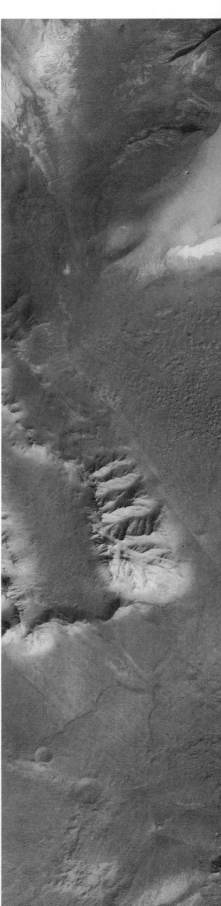

Landslides, faults and V-shaped valleys can be seen within much of Valles Marineris (above). In Candor Chasma (right) there are wind-carved grooves and possible sedimentary deposits formed from ancient lakes.

It is difficult to estimate the age of the canyons, as surviving craters are scarce: many that were formerly present have been removed by erosion. It is also clear that the canyons have been extended by wind and water erosion and by faulting. But it is impossible to determine which of these processes has been the most powerful factor, and how they have altered over time. Herein lies another unsolved problem: what happened to the material that was removed from the surface when the canyons formed?

Overview of the canyons

The Valles Marineris system falls roughly into three main sections. Starting at its western end and moving eastwards, we encounter: Noctis Labyrinthus, a complex maze of interconnected canyons; the central region, consisting of several individual canyons running east-west; and finally chaotic terrain, consisting of irregular canyons merging with each other at the eastern edge.

Noctis Labyrinthus lies close to the summit of the Tharsis bulge. It consists of a sprawling web of interlocking collapsed depressions. These canyons are relatively short and narrow.

The labyrinth is connected to two long, narrow canyons: Tithonium Chasma in the south and Ius Chasma in the north. These most westerly canyons have steep walls, which have collapsed in places, and a number of smaller tributary canyons. Both these canyons expand eastwards and finally merge with the broad Melas Chasma.

Further east, the Valles Marineris system attains its greatest width, being made up of three major canyons running parallel and separated by ridges. Each is about 120 miles (200 kilometres) wide.

To the north is a completely separate, enclosed depression called Hebes Chasma. On its western edge is a small canyon

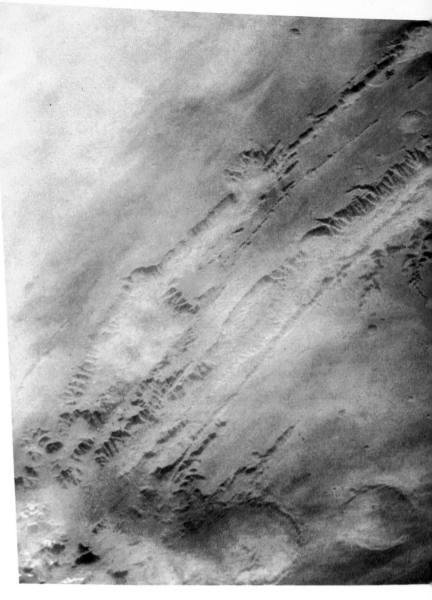

named Echus Chasma, which opens out onto the northern plains. Here the Kasei Vallis channel begins—a proposed manned landing site (see page 84).

The three central canyons broaden and join to form a linear trough known as Coprates Chasma, crossed by many crater chains and faults. Lying west of this is a complicated region made up of canyons and chaotic terrain from which channels emanate.

The canyon walls

The walls of the individual canyons are not consistent with each other, nor are they systematic in their appearance and occurrence. Spurs and gullies are observed as alternate ridges and short alleys. This sort of geological feature probably has the same origin as similar structures seen in

Antarctica, where moderately resistant, uniformly thick rock collapses because of gravitational stresses in the crust.

Perhaps the most spectacular features of the canyon walls are the 'tributary canyons', which have a V-shaped cross-section and are most prominent on the southern edge of Ius Chasma. They have blunt edges with little or no flat floor. Many look like terrestrial 'hanging valleys', with floors higher than that of the main canyon. They were probably formed by water erosion or by 'mass wasting', a process in which the surface collapses under its own weight, but the relative importance of these processes is not known. It is certain that the erosion continued in these valleys after the canyon was first formed, thereby increasing its size.

Left: The canyons' central section, near the region where it attains its greatest width. Right: Tithonium Chasma, a thin, narrow canyon, is separated from the wider Ius Chasma by a plateau. A long crater chain runs across the plateau, parallel to the canyons. Faint mist can be seen throughout the canyon system. Below: The complex structure of Noctis Labyrinthus, consisting of depressions and faults in an intricate network.

The canyon floors

The floors of the canyons vary from place to place. Where a canyon is wider, the floors are flatter: where it is narrow the floors tend to be rougher and segmented. Nevertheless, the canyon floors are very rough when compared to the surrounding plateaux. There are small hills and ridges on the canyon floors, sometimes lightly peppered with craters.

Vast landslides have spilled out onto the canyon floor in many places, especially in Ius Chasma, where the largest is around 60 miles (100 kilometres) long. They have the appearance of being composed of two parts or lobes: a rough inner lobe surrounded by a smoother, broader one. The inner lobes are usually crossed by ridges and valleys, and are probably the remnants of surface rock. The outer lobes consist of what was formerly subsurface rock, more finely broken up.

The presence of subsurface water ice may have been important in these landslides. The canyons are probably deeper than the permafrost layer, so if the ice melted their walls they would have been susceptible to collapse.

Thick, layered deposits of rock are visible in the walls and floors of several canyons, particularly in Ophir Chasma and Candor Chasma. In some regions, the detail and regularity of the layering are remarkable. How were these layers formed? The answer could provide a unique glimpse into the past history of Mars. Some scientists believe that they were caused as the amount of dust deposited by winds changed with variations in climate. Another theory suggests that they are the result of lakes that dried up, leaving behind sedimentary layers.

ECHUS CHASMA

HEBES CHASMA

OPHIR CHASMA

JUVENTAE CHASMA

GANGES CHASMA

TITHONIUM CHASMA

CANDOR CHASMA

NOCTIS LABYRINTHUS

IUS CHASMA

MELAS CHASMA

CAPRI CHASMA

COPRATES CHASMA

EOS CHASMA

THE ATMOSPHERE

The first weather report from another planet was issued the day after Viking 1 had landed on Mars. It read:

Light winds from the east in the late afternoon, changing to light winds from the south-east after midnight. Maximum winds around 15 miles per hour [24 kilometres per hour]. Temperatures ranged from −83°C to −33°C [−117°F to −27°F]. Pressure steady at 7.7 millibars.

The two Viking landers functioned as automatic weather stations for two full Martian years. Their information, coupled with data from the two orbiters, has given us a detailed picture of the weather on Mars. We have learned that the atmosphere of Mars, though much thinner than our own, is very active meteorologically.

Pressures and temperatures

The surface pressure on Mars is less than one per cent of that on Earth: about six millibars, compared to the average terrestrial value of 1013 millibars. As on the Earth, pressure decreases with increasing altitude: higher values are found in low-lying areas and lower ones on mountain-tops. At the lowest point on Mars, in the Hellas basin, the pressure is 8.9 millibars; at the summits of the large volcanoes in the Tharsis region the pressure is as low as one to two millibars. There is no convenient 'sea level' on Mars, so an arbitrary datum level has been chosen, at which the pressure is 6.1 millibars.

Temperatures on Mars are far lower than on the Earth. A typical temperature at the equator is about −60°F (−50°C), though in some equatorial regions the temperature may climb above the freezing point of water and as high as 85°F (30°C) by noon. At the poles a typical winter temperature is around −240°F (−150°C).

The thin Martian atmosphere responds much more rapidly to heating by the Sun than our own atmosphere does. Daily temperature variations are accordingly far larger than those on the Earth, even in our deserts. Typical Viking measurements of the daily temperature range were −22°F to −112°F (−30°C to −80°C); a comparable desert region on Earth, China Lake in California, experiences a range of temperatures from 100°F to 65°F (38°C to 19°C).

However, the times at which the maximum and minimum temperatures are reached are the same on Earth and Mars. This indicates that the prime atmospheric driving mechanism is the same on both planets: convection, the transfer of heat by the movement of masses of air.

Temperatures measured by the second lander at its more northerly location were 10–20°F (5–10°C) lower than those recorded by Viking 1; pressures were greater because it was on lower-lying ground.

Wind patterns changed with the onset of autumn, with maximum wind speeds at around 30 feet per second (10 metres per second). As winter approached, temperatures gradually fell, though this decline was interrupted by the arrival of warmer, global dust storms. These changes were more pronounced at the more northerly latitude of the Viking 2 lander, where the weather was more active.

Cloud types

The Martian atmosphere is mostly composed of carbon dioxide, with minute traces of water, carbon monoxide, and 'inert' (unreactive) gases like krypton and xenon.

An instrument on the Viking orbiters known as the Mars Atmospheric Water Detector (MAWD), measured the amounts of water vapour. Quantities of water vapour can be expressed in terms of 'precipitable centimetres'. This is the depth of water that would result if all the vapour in the air above a given area were to be deposited on that area as liquid water.

MAWD found that the amount of water vapour is highly variable,

The Martian atmosphere is nearly always made hazy by dust and mist. Early morning fog is often seen around Noctis Labyrinthus (right) and frost can fill craters overnight (below).

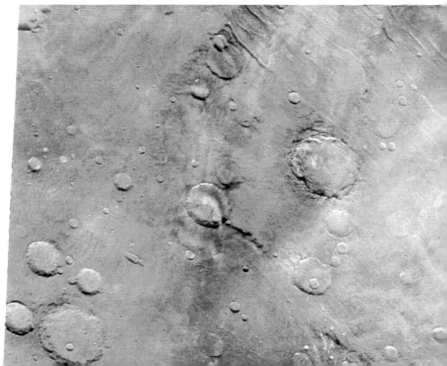

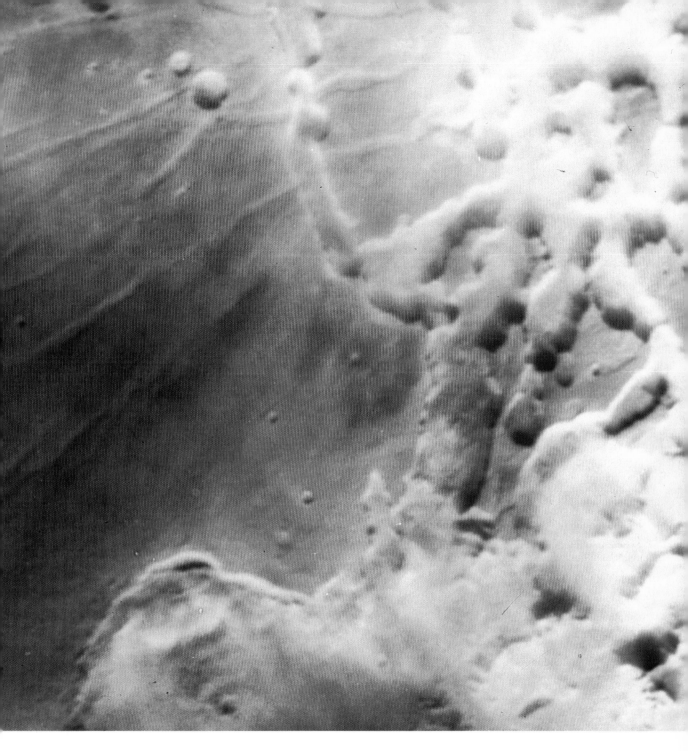

changing from season to season and from place to place. MAWD measurements in midsummer ranged from less than 0.0001 precipitable centimetres in high southern latitudes to 0.01 in high northern latitudes. As the Earth's atmosphere normally contains at least two to three precipitable centimetres, the Martian atmosphere is clearly very dry.

Small though this amount is, it is the maximum that the atmosphere can hold, given its low pressure and temperature. The water vapour is not restricted to low altitudes, but is well mixed in the atmosphere to heights of at least six miles (10 kilometres). Consequently there is plenty of water vapour available to form the clouds that are seen. There are several different types of cloud on Mars, just as on Earth, each having its own method of formation.

Convective clouds are formed when heating of the surface warms the low-lying air, causing it to rise. The air cools with increasing height and has to lose some of the water vapour it contains. Clouds then appear, as distinct 'puffs'. They are most frequently formed in equatorial regions at midday, especially around the slopes of the Tharsis volcanoes.

Wave clouds, looking like the ripple patterns in the wake of a ship, often appear on the lee side of large land masses such as craters, which present obstacles to air flow.

Orographic clouds are formed when air is forced to move slowly and steadily up the slopes of large

The Martian Seasons

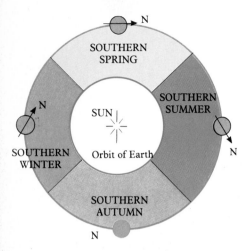

SEASON		DURATION	
Northern Hemisphere	Southern Hemisphere	Martian Days	Terrestrial Days
SPRING	AUTUMN	194	199
SUMMER	WINTER	178	183
AUTUMN	SPRING	143	147
WINTER	SUMMER	154	158

features. The orbiters often saw them in the spring and summer around the Tharsis volcanoes. Early morning fogs also appear in canyons and basins.

Winter temperatures at high altitudes over the poles can be sufficiently low for carbon dioxide to form a thin cloud layer. As we have already noted (see page 10), the seasonal changes in the polar caps significantly affect the Martian atmosphere. Around 20 per cent of the atmospheric carbon dioxide is cycled between the polar caps and the atmosphere each winter season, resulting in a corresponding variation in global atmospheric pressure. Most of the carbon dioxide ice in the seasonal polar caps deposited as atmospheric carbon dioxide comes into direct contact with the frigid Martian soil and condenses. It is also possible that snow storms of carbon dioxide ice occur in winter, helping to extend the seasonal polar cap.

Global weather systems

There is a huge temperature difference between the equator and the winter pole. This produces brisk westerly winds and creates intense low-pressure areas, similar to the depressions or cyclonic systems on Earth.

In the summer hemisphere, Martian weather differs greatly from that on Earth. On Mars, more sunlight reaches the summer pole than the equator, unlike the Earth, where the relatively dense and cloudy atmosphere reflects much of the sunlight, so that less of it reaches the ground at the pole than at the equator.

The result is that in the summer hemisphere there is very little weather and light easterly winds prevail. The dominant wind systems are 'tidal', caused by the expansion and contraction of the atmosphere as a result of solar heating throughout the day.

Dust storms

In 1977 35 individual storms were seen. Two of them developed into global storms; unusually, one occurred before, and the other close to, perihelion (the point of closest approach to the Sun). Normally the global storms develop after perihelion.

In 1979 no global dust storm apparently occurred, although a number of local storms were detected. In 1981/2 very large quantities of dust passed over the Viking lander sites, virtually cutting off the sunlight. Unfortunately both orbiters had shut down by then, so it is not known whether the dust storm was global.

The Vikings showed that global dust storms spread rapidly, eventually enshrouding the whole planet with a featureless haze that lasts

The surface is almost blanked out by a global dust storm—the second of 1977—in this mosaic of pictures taken from orbit. But the faint outline of the Valles Marineris can be seen at top right.

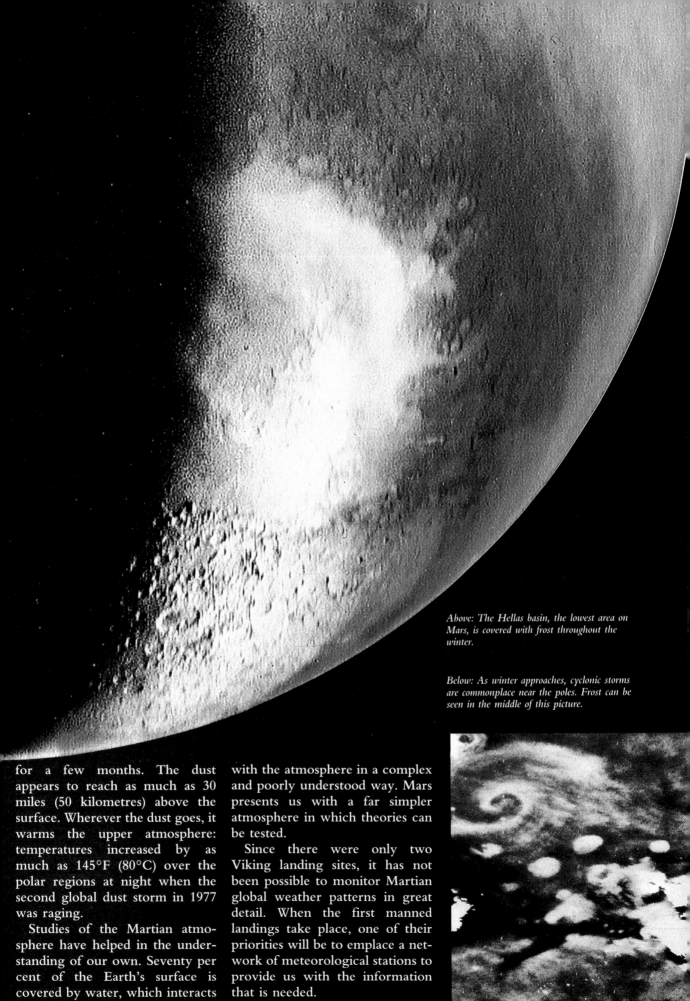

Above: The Hellas basin, the lowest area on Mars, is covered with frost throughout the winter.

Below: As winter approaches, cyclonic storms are commonplace near the poles. Frost can be seen in the middle of this picture.

for a few months. The dust appears to reach as much as 30 miles (50 kilometres) above the surface. Wherever the dust goes, it warms the upper atmosphere: temperatures increased by as much as 145°F (80°C) over the polar regions at night when the second global dust storm in 1977 was raging.

Studies of the Martian atmosphere have helped in the understanding of our own. Seventy per cent of the Earth's surface is covered by water, which interacts with the atmosphere in a complex and poorly understood way. Mars presents us with a far simpler atmosphere in which theories can be tested.

Since there were only two Viking landing sites, it has not been possible to monitor Martian global weather patterns in great detail. When the first manned landings take place, one of their priorities will be to emplace a network of meteorological stations to provide us with the information that is needed.

129

THE ENIGMA OF WATER

Of all the features of the Martian surface, the great channels that scar the surface are among the most unexpected and baffling. These channels have nothing to do with linear markings mistaken for canals by Earth-based astronomers earlier this century (see page 150). If anything, they are even less easy to explain. There is little doubt that they were carved by running water, yet liquid water cannot exist under present conditions on Mars. So what happened to that water?

This is the single most important unanswered question about the red planet. On the one hand the apparent loss of water has important ramifications for studies of climate, not only on Mars but on the Earth too. On the other, knowledge of the present location of water on Mars could affect the planning of manned missions.

The variety of channels

To avoid the problems posed by the idea that the channels were carved by water, it was at first suggested that erosion by lava or the wind created them. These alternatives are no longer seriously entertained. Many of the channels are found on downward slopes, which counts against the wind theory. And some have great systems of tributary branches: it is difficult to see how these could have been formed by wind or lava.

Geologists have distinguished three types of channel on Mars: large, medium and small. They were formed in different ways.

The largest channels are spectacular. They run for thousands of miles, and some are nearly 60 miles (100 kilometres) wide in places. They are found in the equatorial regions, many originating near the chaotic terrain at the eastern end of the Valles Marineris (see page 124). These flow into the Chryse basin, where the first Viking landed, and out across the northern plains. Superficially the large channels look like dried-up river beds on Earth, but geologists prefer to class them as 'outflow channels', as they seem to have originated by catastrophic flooding from below ground. They have few tributaries, and there is much evidence for scouring of the channel floors by debris swept along by the flood. Terraces and teardrop-shaped 'islands' are found downstream in the Chryse region.

But the size of these Martian channels indicates that truly enormous quantities of water were released. It has been estimated that the maximum discharge of floodwater in these channels was between 100 and 10,000 times the average discharge rate of the Amazon today.

Perhaps the source of this water was subsurface ice. Volcanic heat could have melted it, causing the rock above to collapse and release the water in vast amounts.

Medium-sized channels

On the margins between the ancient cratered terrain in the south and the volcanic plains to the north are the medium-sized chan-

A 'fretted' medium-sized channel, about 10 miles (16km) wide and 1.3 miles (2km) deep. The channel is sinuous and landslides can be seen within its walls.

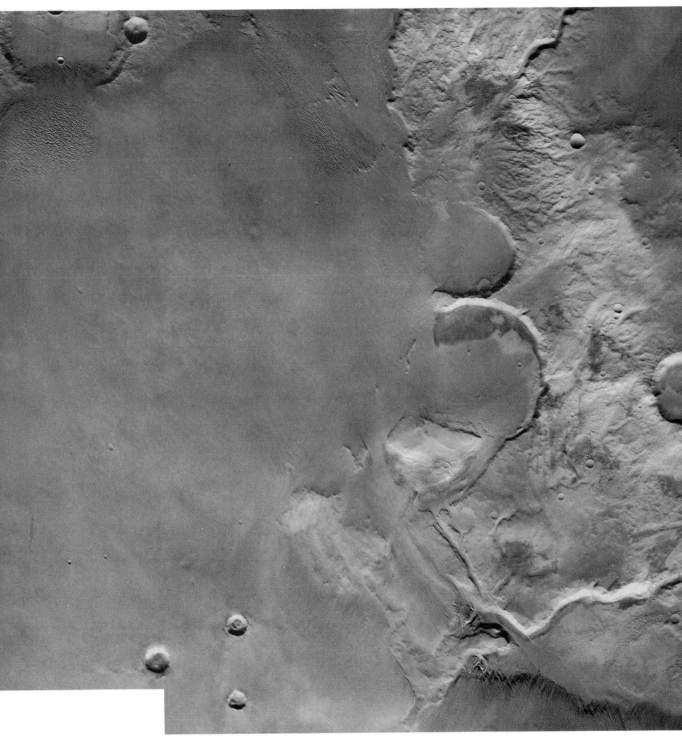

Mangala Vallis lies on the boundary between the planet's heavily cratered terrain and the volcanic plains.

nels. They tend to be sinuous and are joined by many tributaries, increasing in size downstream. Their floors have a curious braided appearance. The channels look like the stream beds that are carved in terrestrial deserts by flash floods caused by sudden rainstorms; but on Mars rainfall seems an unlikely origin, because the atmosphere has never been dense enough to support rain-bearing clouds.

The tributaries often have a characteristic 'blunt-edged' appearance, suggesting that 'spring-sapping' took place: springs appeared at the base of the main stream's cliffs, and eventually caused a collapse that marked the beginning of a tributary. The process was repeated at the head of the tributary and the channel grew longer.

But some geologists believe that

glaciers may also have played an important role in forming some medium-sized channels. This would account for the terrain's curiously patterned ('fretted') appearance in some areas. As the glaciers advanced and retreated, the surface would have been heavily eroded, and the debris would have been carried away by the meltwaters, giving rise to the fretted appearance.

It is certain that, unlike their larger counterparts, the medium-sized channels were not carved by water released by localized heating. Perhaps they were formed in an epoch when the Martian climate was warmer than it is today, and water and ice could exist on the surface.

Valley networks

The smallest channels are the most numerous. They look rather like the dendritic ('treelike') networks of drainage channels seen on Earth. Typically the Martian networks are two to three miles (three to five kilometres) wide, and extend for anything up to 60 miles (100 kilometres). The individual valleys are very narrow; perhaps the water that carved them restricted its erosion to the narrow region at the floor of the V-shaped channel.

At first glance these valley channels look like terrestrial valleys formed by rainwater flowing downhill, and this originally seemed a probable mechanism for their origin. But detailed Viking views showed that, for example, the heads of the channels are totally different from those of rainfed channels on the Earth, and spring sapping is again in evidence. This led planetary scientists to conclude that they were formed by water emerging from underground and flowing over the surface.

These 'runoff channels' are found in the oldest, most heavily cratered terrain in the southern hemisphere, which implies that they too are ancient. But some

The eastern margins of Mangala Vallis show a number of outflow channels, the result of catastrophic flooding billions of years ago. The area shown is the primary candidate for a landing site.

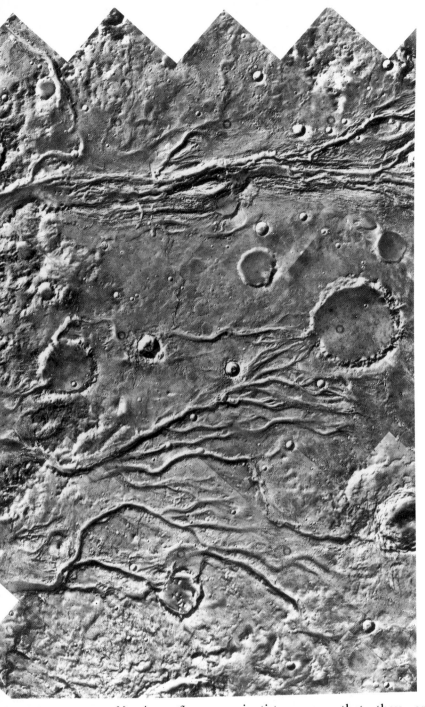

planet's early history: therefore the water would not have evaporated rapidly, and would have had time to carve the valleys.

How old are the other types of channel? The Viking orbiters show that crater counts vary considerably in the floors of larger channels. So they did not all originate at the same time, and range in age from the very young to the very ancient.

Where did the water go?

Unfortunately, we can only guess at how much water Mars originally possessed, how much has escaped from the planet, and how much remains, hidden from view.

The amount of water in the Martian atmosphere today is certainly negligible. The total volume of water vapour in the entire atmosphere is just a quarter of a cubic mile (one cubic kilometre). Ground frosts contain about the same mass of water.

The weak gravitational pull of Mars allows lighter gases to escape into space with relative ease. As the Viking landers hurtled through the upper Martian atmosphere, they observed hydrogen and oxygen—the chemical constituents of water—in the process of being lost to space. The amounts lost are the equivalent of 60,000 gallons (over 270,000 litres) of water per day. This rate of loss may have varied throughout Martian history.

The polar caps are reservoirs of water, but the water ice in them is mixed with frozen carbon dioxide and dust in unknown quantities. Only the permanent ice cap at the north pole shows evidence of being predominantly water ice. The thickness of this cap has been estimated at anything from 30 feet to half a mile (10 metres to one kilometre).

Underground permafrost is another reserve of water. Generally speaking, the colder the conditions, the greater the chance of finding permafrost near the surface. Estimates of subsurface

Many large outflow channels open on to Chryse Planitia, near Viking 1's landing site. Tear-shaped islands (left) were formed when plateaux were eroded by water. The uppermost crater is fresh and so postdates the catastrophic flooding that formed the channel. At the western edge of Chryse (above) a number of branched channels converge. Many of the craters have been cut into by the channel flows, showing that they were formed before the flooding.

scientists argue that they are located there for different reasons: perhaps the older terrain is for some reason more easily eroded; or perhaps there were greater water reserves below the surface in older regions.

Because the valley channels are smaller than the other two types, it is clear that much smaller quantities of water were involved in their formation. This in turn suggests that, if they are indeed ancient (almost certainly more than three billion years old), the atmosphere was thicker in the temperatures on Mars suggest that at the poles the permafrost might well be up to five miles (eight kilometres) thick, and lie only an inch (a few cm) below the surface. At the equator it could be up to two miles (three kilometres) thick, and about 10 feet (a few metres) deep.

The gamma ray spectrometer on NASA's Mars Observer may reveal ice just below the surface. But again, until core samples are drilled, all these figures are hypothetical. A network of seismic soundings made by astronauts would produce a detailed picture of the distribution of permafrost and how far it extends into the Martian crust.

Climatic changes

Because Mars is farther than the Earth from the Sun, it formed under cooler conditions, and planetary scientists would expect more water to have condensed. One recent study suggests that during the first two billion years of Martian evolution some 1.6 million cubic miles (6.7 million cubic kilometres) of water were outgassed by the volcanoes, equivalent to a layer 152 feet (46 metres) deep, covering the globe. Because volcanism has spanned the whole of Martian history, water may have been released continuously, though not at a uniform rate. Water may have also come from meteoritic and cometary impacts.

Other estimates, based on the amount of water believed to be locked in the poles and below the surface, plus that which has been lost to space, suggest that the total quantity of water involved may have been equivalent to a planet-wide layer of water as much as half a mile (one kilometre) deep.

Water may well be chemically combined with the rocks; the Viking landers were not able to split them open for chemical analysis.

The fact that vast quantities of

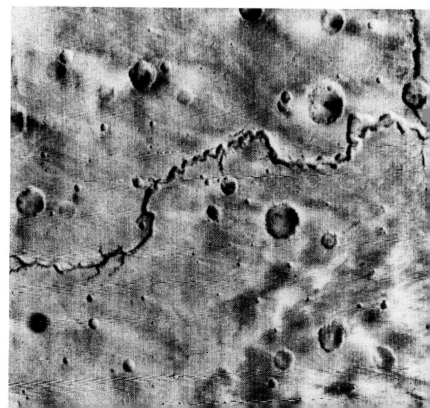

Left: Many of the large channels emerge from the chaotic terrain at the eastern end of Valles Marineris. The source of Tiu Vallis is the roughly circular area about 60 miles (100km) across, just below the centre of the picture.

Below left: Nirgal Vallis is about 480 miles (800km) long, with many tributaries. They are relatively short, which suggests they were formed by underground 'spring sapping'.

Right: Ares Vallis is some 12 miles (20km) wide and a few hundred yards deep. There are very few craters on its floor, suggesting that it is more recent than the surrounding terrain. Windstreaks can be seen at the top of the picture.

water once flowed on Mars means that the atmosphere must then have been denser, or warmer, or both. (Atmospheric pressure is important because water evaporates faster under a low pressure than under a high one.) Originally Viking scientists thought that the Martian atmosphere could have been as dense as our own, but this seems highly unlikely. For water to have flowed as a liquid, the pressure need only have been as high as 30 mb at most, if Mars was

warmer because of changes in its orbit (see page 138). As time went on, Mars progressively lost more and more of its atmosphere into space, until today the surface pressure is less than one hundredth of the Earth's.

The Martian channels were

probably formed at periods when the climate was warming up. Because they have been formed throughout the whole of Martian history, it seems likely that the planet has undergone climatic cycles throughout this time (see page 138).

Our knowledge of the composition of the poles comes from infrared measurements taken by the Viking orbiters. The summer temperature at the permanent north pole is around $-95°F$ $(-70°C)$, close to the temperature at which atmospheric water vapour would form frost in Martian conditions. This suggests the northern cap is for the most part composed of water ice.

Both permanent caps are a few degrees of latitude across, though the northern one is larger than the southern. The amount of light reflected from the surface is much less than that from terrestrial ice fields. This suggests that the Martian ice is dirty—probably mixed with dust, silts and clays.

Temperatures measured at the south pole are substantially lower: around $-165°F$ $(-110°C)$ at the permanent cap during summer. This is nearer to the frost point of carbon dioxide, which is therefore presumed to be the major component of the southern cap. More light is reflected from the permanent southern cap than from the northern, indicating that it is much more free of dust. At its minimum, the remnant cap is centred on a point four degrees from the pole, and its edges just cover the pole.

Ice and dust

Why are the permanent caps different from each other? One theory is that the annual global dust storms are the culprits. When it is summer in the southern hemisphere Mars is closest to the Sun, and the extra heating causes winds that whip up dust storms. The Viking landers observed that as much as 20 per cent of the atmospheric carbon dioxide freezes out onto the winter pole. So this condensation occurs in the northern hemisphere at a time when global dust is present in the

air. Some of this dust is incorporated into the polar cap, remaining well mixed with the ice and discolouring the permanent cap below. During the southern hemisphere's winter much less atmospheric dust is present, and the seasonal cap is 'cleaner'.

Both polar caps have a swirling appearance. The winds, spiralling from the poles, combine with solar heating to remove the intermingled dust and ice from slopes that face the equator. The ice of the southern polar cap shows swirl patterns in a clockwise direction going outwards, while its northern counterpart swirls in a counterclockwise direction.

Valleys and escarpments are visible through the ice. A

prominent valley in the north, named Chasma Boreale, and a similarly broad one at the south pole, called Chasma Australe, have been identified as possible landing sites for future missions.

The seasonal caps

Though the southern hemisphere summer is relatively short and hot, the winters there are longer and colder than in the north (see page 128). The seasonal covering of carbon dioxide ice is therefore much greater. It has been estimated that a layer about 9–20 inches (23–50 centimetres) thick freezes out on the pole at this time. The southern seasonal cap is roughly circular and, unlike the permanent cap, centred on the

This photomosaic of nearly 400 Viking 2 orbiter frames shows the residual north polar cap, composed primarily of water ice.

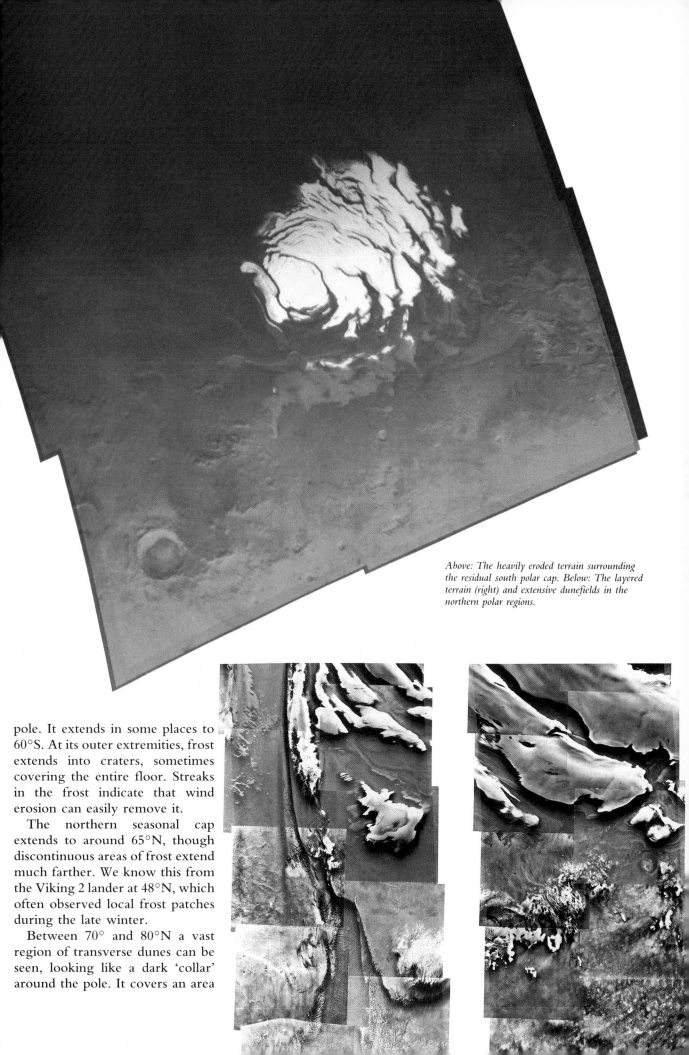

Above: The heavily eroded terrain surrounding the residual south polar cap. Below: The layered terrain (right) and extensive dunefields in the northern polar regions.

pole. It extends in some places to 60°S. At its outer extremities, frost extends into craters, sometimes covering the entire floor. Streaks in the frost indicate that wind erosion can easily remove it.

The northern seasonal cap extends to around 65°N, though discontinuous areas of frost extend much farther. We know this from the Viking 2 lander at 48°N, which often observed local frost patches during the late winter.

Between 70° and 80°N a vast region of transverse dunes can be seen, looking like a dark 'collar' around the pole. It covers an area of more than two million square miles (five million square kilometres)—much more than the combined areas of the Saharan and Arabian deserts. The dunes tend to mask the underlying terrain, though they are broken up by larger surface features. Only localized dunes are seen at the south pole.

The northern dune fields are very perplexing. It is not clear where the sand in the dunes came from or how the dunes persist, despite being repeatedly covered and uncovered by the advance and retreat of the seasonal ice.

The layered terrain

By far the most puzzling of the polar landforms are the thick layered deposits that extend from the poles as far as 80° latitude. They consist of mixed dust and ice, the proportions of each varying from one layer to the next.

At the north pole, where the residual cap is larger, the layered terrain is covered by ice all year round. The ice deposits here are between 2.5 and 3.5 miles (four and six kilometres) thick, while at the south pole they are between half a mile and one mile (one and two kilometres) thick. The layering within the ice appears most prominently in the form of fine horizontal bands on slopes from which the frost has been removed by wind and the warmth of the Sun. The layers are remarkable for their regularity and extent. They range from 30 to 160 feet (10 to 50 metres) thick, and some run for hundreds of miles.

The overall appearance of the layered regions is of smooth, gently undulating terrain into which the valleys and scarps cut. Very few craters are seen in these regions, suggesting that the terrain is young and therefore preserves a record of recent Martian history.

The layering probably occurs because of variations in the proportions of dust and ice deposited by the winds. Its regularity

suggests that conditions on Mars have varied in a regular cyclic manner in the past. Why should the Martian climate have changed in this way?

The simplest answer is that changes in the planet's orbital and rotational movements were the cause. The movement of Mars has been altered over time by the gravitational influence of the other planets (particularly Jupiter), and by the non-uniform distribution of mass within its own crust.

The result is that, firstly, the *eccentricity* of the orbit of Mars varies. The orbit is distinctly ellip-

The vast dunefield surrounding the north pole appears as a dark band at the lower left of this picture. Layered terrain can be seen at top right.

tical, and its eccentricity is the amount by which it departs from a perfect circle. This changes over a period of 100,000 years.

When the eccentricity increases there will be an increase in the amount of sunlight falling on the hemisphere that experiences summer when Mars is closest to the Sun. At present it happens to be the southern hemisphere that thus receives more sunlight during the summer, and the dust that is

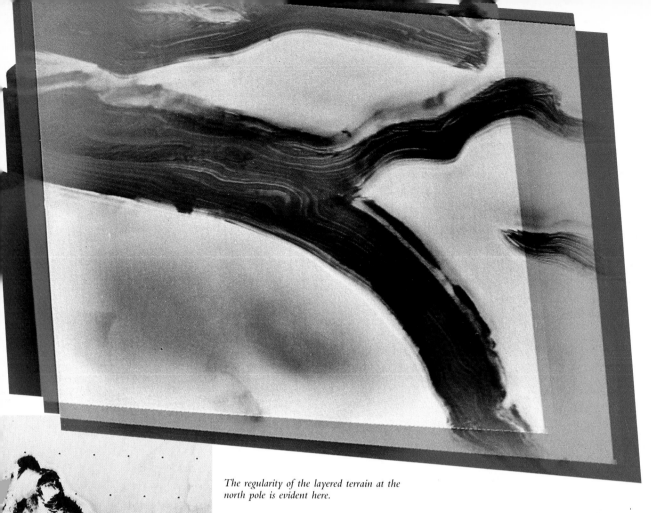

The regularity of the layered terrain at the north pole is evident here.

Great ice cliffs are seen within the north polar icefields.

raised in the global dust storms is deposited on the northern pole.

Secondly, Mars's *obliquity* also changes over millennia. Obliquity is the tilt of the planet's axis in relation to its orbit about the Sun, and it is responsible for the seasons. The current value is 24°, but the maximum tilt could have been as high as 46° within the first half-billion years of Martian history. However, the uplifting of the Tharsis bulge probably reduced this maximum figure to 35°. At present, the obliquity of Mars probably changes by as much as 10° over a period of 100,000 to a million years.

When the obliquity is greater towards the Sun, the summer pole would experience relatively higher temperatures: less carbon dioxide would freeze out on the winter pole and atmospheric pressure would tend to stay higher. Slower-moving winds would then be sufficient to raise dust, so that more dust would be deposited at the poles.

When the obliquity is less, the temperature of the summer pole would be lower: more ice, but less dust, would be deposited. In this way the layering of the terrain could come about.

Finally, planets also *precess*, or wobble around their axes, like spinning tops: one cycle of Martian precession takes 175,000 years. The pole on which the deposition occurs may alter throughout each precession cycle.

The combined effects of these changes could have affected the Martian climate considerably. During the warmer periods, liquid water could have flowed across the Martian surface. There might even have been seas and lakes in these eras. There are sediments in the canyons of the Valles Marineris system, suggesting this (see page 125).

A core sample drilled out of the layered deposits at the poles could be dated by the natural radio-activity of the substances within it. This would reveal when the Martian climate was hotter, when cooler. These results could be compared with dates determined from seabed sediments on Earth, which also suggest cyclic climatic changes.

If it were found that cold epochs coincided on Earth and Mars, some common factor, such as changes in the heat output of the Sun, might be their cause. Otherwise, orbital changes—which occur independently for the two planets—would be indicated. In either event, the comparison could answer one of the most puzzling questions about the Earth: how ice ages are triggered. Thus study of Mars could be of direct relevance to understanding our planet's past.

THE MOONS

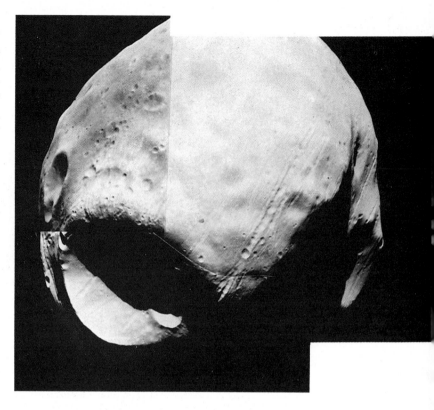

In the summer of 1877, the American astronomer Asaph Hall spent exasperating weeks searching for satellites of Mars. The story goes that he was so discouraged that his wife Angelina repeatedly had to order him back to the telescope.

Her firmness paid off: in August, Hall discovered two tiny worlds circling Mars. He named them Phobos (fear) and Deimos (panic), after the two acolytes of Ares, who was the Greek god of war and the equivalent of the Roman god Mars.

Nearly a century later, Angelina Hall was honoured for her part in the discovery: a crater that dominates Phobos, first seen in the pictures sent back by Mariner 9, was given her maiden name, Stickney.

Paths of the moons

The orbits of both moons are circular, lying directly above the planet's equator. Phobos, the larger of the two, is the nearer, moving 5827 miles (9378 kilometres) from the centre of Mars. It is the only moon in the solar system to circle its parent planet in less time (7 hours 39 minutes) than the planet takes to revolve on its own axis (24 hours 37 minutes). Therefore, as we noted previously (see page 93), it rises in the west as seen from the surface of Mars (though it is so low above the surface that it cannot be seen at all from the polar regions beyond 70° of latitude).

Deimos is much farther out, some 14,577 miles (23,459 kilometres) from the centre of Mars, and takes 30 hours 18 minutes to complete one orbit.

The rotation of Phobos and Deimos is 'locked onto' their parent planet, so that they turn on their axes in the same time that they take to orbit Mars. The result is that each presents one face to Mars all the time—just as our Moon presents only one face to us.

Both Phobos and Deimos move in orbits on the borderline of instability. Phobos is close to the 'Roche limit', below which the moon would break up. Deimos is close to the point beyond which the gravitational pull of Mars is so weak that the moon would gradually escape into space.

The orbit of Phobos is slowly 'decaying': the satellite is spiralling in close to Mars, speeding up as it does so. This happens to artificial satellites orbiting the Earth: they enter the Earth's atmosphere and plunge to a fiery death (as the Skylab space station did in 1979). However, their demise is hastened by the drag of air resistance. The timescale for the destruction of Phobos is much greater: it will be a hundred million years before it either breaks up from stresses or crashes into the planet.

Deimos, on the other hand, seems to be getting further and further from Mars, slowing down as it does so, and will eventually head off into space.

The crater Stickney (seen partly in shadow) dominates the hemisphere of Phobos which always faces Mars.

Born from the asteroids?

What are the moons made of? Until the Soviet Phobos craft drop landers and analyse the surface material directly, scientists have to make the best guesses they can from the Viking observations.

One clue lies in the way the Martian moons reflect light. They are among the darkest objects in the solar system, reflecting only six per cent of the sunlight that falls on them. They are, in fact, darker than coal.

Another clue is provided by their density. Earth-based observers, monitoring the orbits of the moons, long ago estimated that they were very light in relation to their size. In 1960 the noted Soviet theorist Iosif Shklovskii caused a stir when he suggested—jokingly, as he later admitted—that they might be hollow.

The Viking orbiters provided the opportunity to measure the moons' densities accurately. The

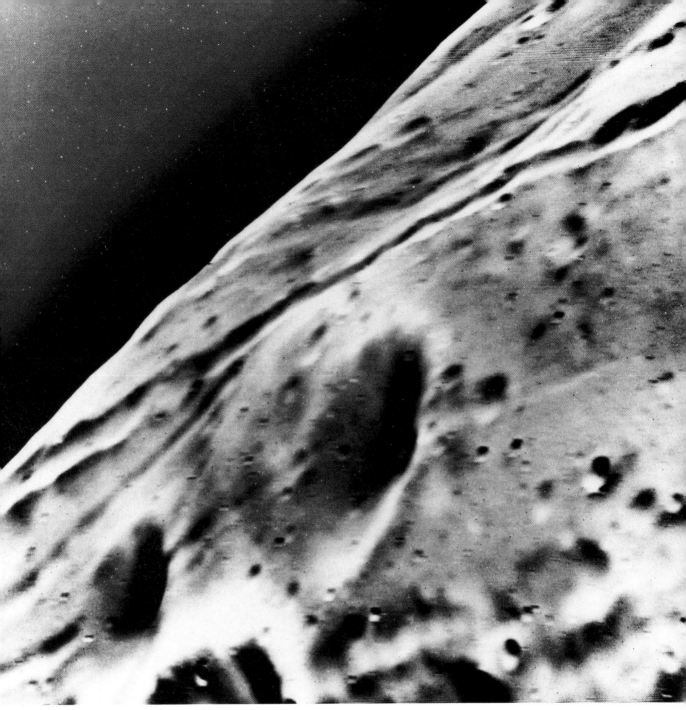

gravitational pull of the satellites on the craft as they passed very close revealed the mass of each moon. It turned out that they have a density about twice that of water—half Mars's density. One Viking scientist said that they were 'more marshmallow than rock'.

The moons' low density and dark colour, together with Viking measurements of how they reflect infrared light, suggest that they are similar to one type of meteorite, the carbonaceous chondrites. These are believed to be fragments of 'C-type' asteroids, which belong to the outermost part of the asteroid belt. This suggests that Phobos and Deimos are captured asteroids.

But there are doubts about this. If they were captured bodies, their original orbits would have been randomly related to Mars: they would probably have been non-circular, and tilted with respect to the Martian equator. It is unclear why such orbits should have changed to the circular equatorial orbits we see now. Tidal forces, which are responsible for the fact that each satellite keeps one face always turned towards Mars, are a possible explanation, but the moons are so small that it is doubtful that the tidal forces would be strong enough.

But if, on the other hand, Phobos and Deimos were formed in the vicinity of Mars from material left over after the planet's birth, why are the satellites so different in composition from the mother planet?

Mysterious grooves

The two moons are about equally densely cratered, suggesting that their surfaces are equally old—about three billion years.

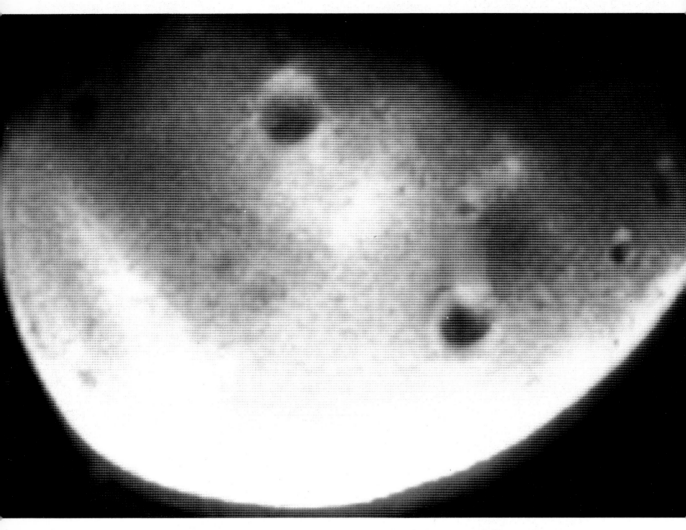

Deimos, the smaller of the moons, is uniformly grey: colour variations are the product of image processing.

Phobos measures about 17 by 14 by 12 miles (28 by 23 by 20 kilometres). It is peppered with craters, and looks like the highlands of our Moon. Crater Stickney, six miles (10 kilometres) across, dominates the surface; nearby are two craters about three miles (five kilometres) across.

Ejected material, excavated when the craters were formed, fell back onto the surface. It is surprising that Phobos clawed back so much of it, since the escape velocity of the tiny world is only 50 feet per second (15 metres per second). However, the ejecta certainly fell back slowly, for it failed to create secondary craters.

Clearly marked grooves radiate from Stickney in all directions, converging on the opposite side of the moon. Close to Stickney they are 2500 feet (700 metres) across and 300 feet (90 metres) deep. With increasing distance from the crater their size decreases until on the other side of Phobos they are less than 350 feet (100 metres) wide.

The grooves certainly look as if they were created by the impact that formed Stickney, which almost shattered the moon. Crater densities in the grooves are similar to those on the surrounding terrain: if the grooves had been created throughout the history of Phobos, there would have been fewer craters in the younger ones.

But some scientists feel that the peculiarities of the orbit of Phobos may have played a part in producing the grooves. Stresses caused by Martian gravity and Phobos's slow inward spiral towards the planet could have fractured the surface. But this would still leave the grooves' convergence on Stickney to be explained.

In addition, planetary scientists think they can explain the characteristic features of the grooves on the assumption that they were formed by just such an ancient impact. The material of Phobos probably contained 10–20 per cent of water—this is the proportion of water found in carbonaceous chondrites. The water would have vaporized when the impact occurred and helped to form the raised rims that characterize these grooves.

The terrain of Deimos

Deimos measures 10 by 7 by 6 miles (16 by 12 by 10 kilometres). Though it is also heavily cratered, the largest crater that is recognizable is only 1.4 miles (2.3

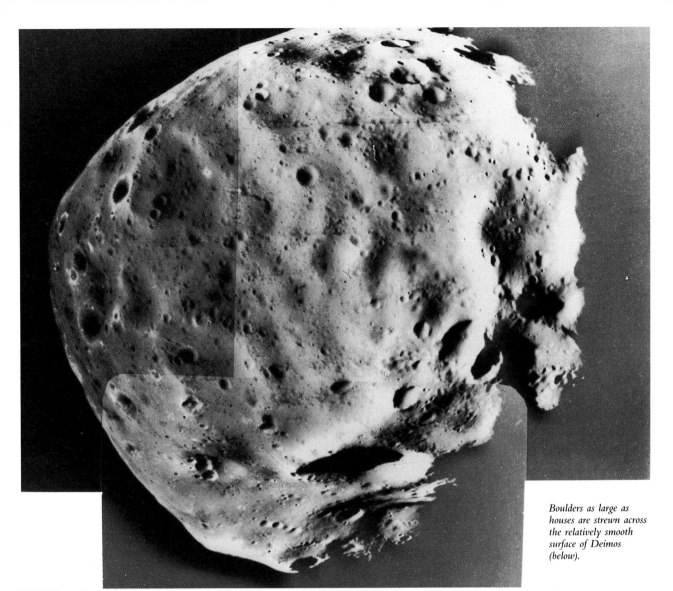

Boulders as large as houses are strewn across the relatively smooth surface of Deimos (below).

This Viking orbiter mosaic (above) shows that the heavily cratered surface of Phobos resembles the highlands of our own Moon.

kilometres) across. The surface relief is more muted than on Phobos: most of the craters are partially filled, and are noticeable only by the relative brightness of their outer rims. The depth of the layer within Deimos's craters has been estimated at 15–35 feet (5–10 metres). It is not clear why Deimos should be so different from Phobos, where craters range from the sharply marked to the heavily degraded.

Studying the moons of Mars will not be as glamorous as exploring the red planet. But they have their own share of scientific curiosities to be resolved. As targets that are easier to reach than the surface of Mars, and easier to depart from, they could prove an invaluable bridgehead to the world below.

PART FIVE

EXPLORING

The first Mars mission will be far more sophisticated than the first Moon landing, in which the main goal was to achieve a safe touchdown. Its main aim will be exploratory, searching for useful raw materials such as water, while building up a comprehensive scientific picture of the planet—its current state, its history and its future development.

At the moment, for example, we know very little of the planet's internal structure. The Viking landers provided only very basic information, and the unmanned missions in the 1990s will not be able to match the wealth of scientific data that can be generated by a manned expedition. Astronauts are much more adaptable than robot vehicles: they are able to assess a wide variety of situations very rapidly and adjust their actions accordingly. A geologist with a rock hammer can collect more useful samples systematically in an hour than an automated rover could in a year.

Human explorers will be especially valuable in the search for life. A trained astronaut can easily spot protected sites that would be favourable to life, such as a spot sheltered by a rock. Automatic equipment will be better employed on repetitive measurements stretching over long periods.

Scientific research will fall into two broad categories: day-to-day exploration by the astronauts, concentrating on the more difficult tasks, and long-term monitoring by automatic packages left behind when the human team departs.

Automatic science stations will be systematically placed around the lander to continue observations after the astronauts have departed.

Astronauts on Mars, like their predecessors on the Moon, will photograph the surface in detail (see gnomon in mid-picture) before collecting samples.

Geological investigation will be the major concern. At least one geologist will be included in the crew. Geology teams at mission control will plan exploration traverses, using the orbiter's high-resolution pictures of the surface. They will transmit facsimiles of route maps that they have worked out to the exploration crew. The astronauts will not only need to

collect samples to bring home but also to study rocks *in situ* and to conduct some analyses on board the lander.

Looking at the chemical and mineral compositions of the rocks will provide information on the geological history of Mars—where the rocks were formed and at what temperatures and pressures, the development of the atmosphere, the history of water on the planet. Much information will be gained from rocks visible at the surface, but drilling deep into the ground and removing rock 'cores' will reveal the historical development of a region's geology.

All in a day's work

Each day the astronauts will set out on their rover, carrying a set of standard geological tools: hammers, chisels, rakes, sieves, tongs—to enable them to pick up rocks despite their cumbersome spacesuits—sample bags, weighing scales, magnifying glasses for preliminary assessments, and the ubiquitous gnomon. This is a tripod with a free-hanging central rod, which is photographed against each sample before it is removed from its location. It shows the scale of the objects photographed and the slope of the ground, and its shadow indicates the direction of the Sun. A colour pattern attached to the gnomon will allow scientists back on Earth to reconstruct the surrounding colours precisely. This is essential because cameras always falsify colour to some extent.

After the samples have been photographed and their locations carefully documented, they will be placed in hermetically sealed containers. Geologists recommend that at least a quarter of the samples should be kept refrigerated at Marslike temperatures to prevent changes *en route* to Earth.

The rover will stop every few

To drill shot holes for seismic exploration, a percussion drill similar to the jack-hammers used in terrestrial mining will be employed (above). The drill will be powered by a compressor using atmospheric carbon dioxide. It is estimated that a hole 10ft (3m) and 2in (5cm) in diameter could be cut in half an hour or less, depending on the soil type. The drill rod is hollow and rotates (right), which increases its cutting ability. Loose chips of rock are blown away, out of the hole. Shallow core samples would require a more complex drill.

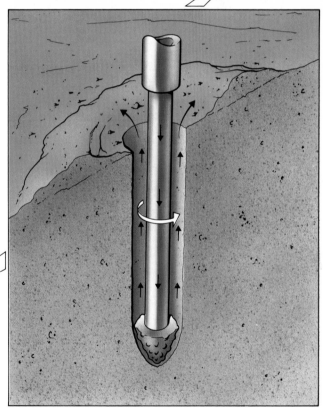

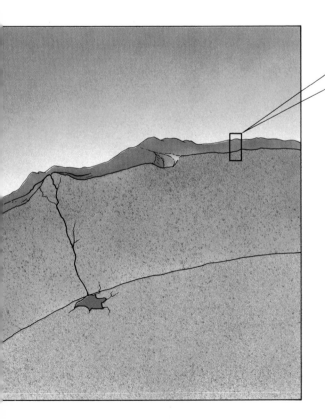

Controlled seismic explosions (left) will reveal much about the internal structure of Mars, which is little understood at present. The Martian crust is rigid and around 30 miles (50km) thick. The mantle is probably cooler and thicker than the Earth's. The core may be from 1,600 to 2,600 miles (2,600–4,200km) in diameter—the uncertainty is due to our ignorance of its temperature.

The surface explorers will train around terrestrial craters, such as Meteor Crater in Arizona, to learn geological investigation techniques.

hundred feet, and the soil will be tested automatically. Its mechanical strength will be measured by penetration tests, in which probes are thrust into the soil, and by weight-bearing tests, in which a metal plate is pressed onto the surface with a known force. The astronauts may also crush rocks in a calibrated handheld device.

Drilling for samples will be carried out regularly. Apollo astronauts found this to be one of their most strenuous activities. The rover will carry electrically powered drills and a supply of aluminium tube sections.

Probing Mars

The internal structure of Mars will be better understood when the rate of heat flow from the interior has been studied. This can be done with thermometers placed at different depths; the holes left after drilling cores will be useful for this. Small heaters placed near the thermometers would reveal how heat flows through the material making up the surface layers.

Seismometers will also be buried to register Marsquakes. They will record tremors with frequencies ranging from one vibration every 10 seconds to 50 vibrations per second. These will reveal much about the internal structure of the planet, but at least three widely separated sites would be required —preferably thousands of miles apart—to pinpoint the origin of a quake. Miniature artificial Marsquakes could be created, by means of small explosive charges, and perhaps even by crashing discarded spacecraft stages onto the surface—subject to the requirement that the planet is not biologically contaminated. Devices to measure local variations in the gravitational and magnetic fields will also provide clues to the Martian interior.

There will probably also be complex equipment for detailed

Left: Trained astronauts will be able to choose which of the myriad surface rocks are scientifically the most interesting.

Right: The colour gradations on the gnomon seen in the foreground of this Apollo 16 photograph will be useful in calibrating surface colours on Mars.

A soil sifter, one of the variety of instruments that will be used to probe surface material.

soil analyses. A scanning electron microscope can be used to look for tiny fossils and aid in identifying minerals. An X-ray fluorescence spectrometer, which records the X-rays emitted by materials when irradiated by a radioactive source, will determine the elements present. A combined gas chromatograph and mass spectrometer will be used to separate and measure the gases driven off when a sample is heated.

When rovers and eventually a manned expedition reach the poles, core samples taken from the layered deposits of ice and dust (which are hundreds of millions of years old) will be the first priority. These will constitute a record of the climatic history of the planet, just as the ancient ice of Antarctica does on the Earth.

Science stations

Several automatic 'science stations' will be established on the surface, powered by small nuclear generators with a lifetime of as much as 10 years. Their instruments will probably include magnetometers, seismometers, heat probes and meteorological instruments. The latter will measure the pressure, humidity and temperature of the air, the speed and direction of the wind, and the amounts of windborne dust. During the manned phase of the mission, they will help to predict weather conditions for the explorers. A mass spectrometer will monitor the composition of the air.

When the astronauts blast off for Earth, the mission will be far from over. The automatic science stations will continue their work tirelessly, beaming their data to an orbiting relay satellite, or directly to the Earth. As on the Apollo missions, a central unit will take readings from several outlying instruments and retransmit them, in addition to relaying commands from the Earth.

Despite centuries of telescopic observation, followed by intensive scrutiny by spacecraft, there is still no clear-cut answer to the tantalizing question: 'Is there life on Mars?' The results of the Viking lander biological experiments were inconclusive. It could well take a manned expedition to end the debate once and for all.

Canals and controversy

When dark areas were first seen on Mars by telescopic observers, they were thought to be vegetation or seas. We now know that they are simply regions of darker surface rock. More controversial were linear markings first reported by Giovanni Schiaparelli, director of the Milan Observatory, in 1877. Schiaparelli termed the markings *canali*, meaning 'channels' or 'grooves', but this was almost inevitably rendered as 'canals' in English—with the clear implication that they were artificial.

Many observers saw no such markings. But one man who did became a vociferous supporter of the idea that they were artificial. Percival Lowell set up a private observatory in Flagstaff, Arizona, and during the 1890s and early 1900s observed literally hundreds of canals. He became convinced that they were irrigation networks, built by a highly advanced Martian civilization desperately trying to overcome arid conditions.

But their very existence was vehemently denied by many scientists. We now know the canals are spurious: they were most probably the result of the brain's tendency to join barely perceivable features into lines. Curiously enough, Lowell is said to have had the ability to recognize telegraph wires in the Arizona desert at a distance of a few miles: an optician said he had the keenest eyes he had ever examined. But superimposing

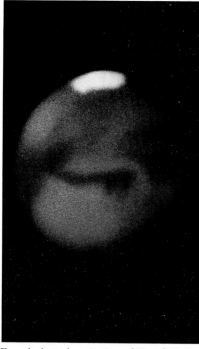

Even the best telescopic views of Mars from the Earth reveal only colour variations, such as the bright poles and the dark expanses previously mistaken for vegetation.

From this view of a partly shadowed, wind-eroded rock has arisen the myth of 'the Face of Mars'.

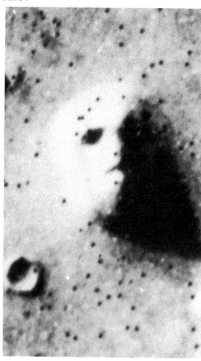

Opposite: The notorious 'B block' (left, mid-picture) observed by the first Viking lander, on which the shadows vaguely resemble the letter B.

Lowell's canal networks on modern spacecraft maps shows that they bear no relation to any real surface features.

Nevertheless the notion of canals and intelligent Martians coloured studies of Mars for most of this century. When in 1900 a French newspaper offered a prize of 100,000 francs for contact with extraterrestrial life, it excluded Mars on the grounds that it would be too easy.

The sad fact remained that Mars was tantalizingly elusive through even the best telescopes. It was clear that probes would have to be sent there if any better understanding was to be gained. When Mariner 4 flew by Mars and sent back 11 pictures showing craters, it virtually put an end to the notion of a Lowellian Mars. The red planet became the dead planet in the popular imagination. More damaging to the possibility of life than the craters was Mariner 4's revelation that the Martian atmosphere was much thinner than previously supposed, and mainly composed of carbon dioxide. In fact measurements by the next spacecraft in the series, Mariners 6 and 7, suggested that it was nearly 99 per cent carbon dioxide. There was clearly little chance that any life forms more advanced than microbes could exist on Mars.

If such micro-organisms *did* exist, they would have evolved slowly, changing little over the millennia, since they would almost certainly be dormant for most of the time to avoid extremes of cold and aridity. Mariner 9's observations of extensive dried-up channels clearly implied that there had been supplies of water on Mars in the past. Water is essential for most biological processes on the Earth, playing a role as important as that of DNA at cellular level. But present-day atmospheric limitations on liquid water on Mars severely restrict the possibilities of life surviving there.

The Viking invasion

It was clear that the next step forward in the search for life on Mars would be the direct analysis of surface soil. This was the primary aim of the Viking landers. But tight constraints on space and power severely limited the range of experiments each lander could carry. In the end the designers managed to fit a fully automated biological laboratory—probably the most complicated machine ever built—into a volume of only two thirds of a cubic foot (less than a fifth of a cubic metre).

It worked with soil samples collected by an arm that extended to a maximum distance of 10 feet (three metres) from the body of the lander.

When the experiments were tested in a variety of environments on the Earth, including Antarctica, the biology instruments detected life in every case. But the Viking instruments could detect Martian life only if it was carbon-based, like

Engineers putting the final touches to the automated biological laboratory that formed part of the Viking lander's scientific payload.

terrestrial life. The question was: would Martian microbes, if they existed, share this same biochemistry?

New myths about Mars

In view of the immense popular interest in Mars, it is understandable that myths should have sprung up around the Viking project even before the landers had touched down.

Just after Viking 1 arrived in orbit, an Italian writer claimed that it had discovered an ancient city,

but that the pictures had been censored. How he came by this information in Milan has never been explained. Later in the mission, the first orbiter returned a picture of a wind-carved hill, a mile (1.5 kilometres) high, on which the shadows vaguely resembled a human face. It was christened 'the Face of Mars', and is still cited regularly as evidence of a fully fledged Martian civilization.

One picture taken by the Viking 1 lander just before sunset reveals a marking curiously like the letter B on a rock face, the product of wind erosion and the lighting conditions. Viking scientists had often warned that erosion could produce curiously shaped rocks but they were not heeded.

A myth has arisen over pictures of another rock which seemed to be more greenish in later pictures than earlier ones. The change was due to dust deposits—but it has been attributed to the growth of organisms resembling lichen.

The Viking results

The biology experiments on the

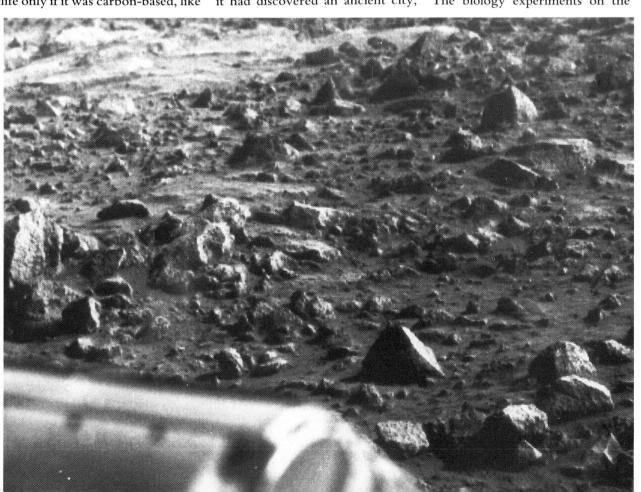

two landers searched for micro-organisms.

Three experiments, based on different sets of assumptions, formed the Viking biology package.

The *Pyrolytic Release* (PR) experiment assumed that Martian microbes, like those on Earth, would assimilate carbon dioxide and produce carbon from it. The sample was illuminated by a xenon lamp to simulate the Martian sunlight and was surrounded with carbon dioxide to simulate the Martian atmosphere. The gas contained a trace of radioactive carbon-14. The gas was removed and the sample was heated to drive off any radioactive carbon taken up by microbes. Some was indeed detected but, because the experiment could be run only once, no firm conclusions could be reached.

The *Gas Exchange* experiment (GEX) assumed that Martian microbes would be used to Earth-like conditions. An inorganic 'soup' of nutrients was mixed with the soil sample to reveal any evidence for metabolic activity. Because the sample was dry, it was slowly humidified: during this time it rapidly gave off oxygen. But when the soup was added, there was no reaction. Experimenters concluded that the soil contained a very powerful oxidizing agent but that no life form had been detected.

The *Labelled Release* (LR) experiment was similar except that the nutrient soup was laced with radioactive carbon-14. Again the hope was that any microbes would feed on the soup and give off tell-tale radioactive carbon dioxide. This was indeed detected, but the flow ended abruptly before the soup was used up. Terrestrial organisms would have produced carbon dioxide for as long as any nutrient remained. Again, a chemical reaction not involving life was suspected.

So the biological experiments taken alone were at best ambigu-ous. But the most damning piece of evidence came from a separate instrument called the Gas Chromatograph / Mass Spectrometer (GCMS). Its purpose was twofold: soil samples were heated to break them down into their constituents, which were detected by a mass spectrometer; and the composition of the Martian atmosphere was analysed by a gas chromatograph. The instrument could detect concentrations as low as one part in a million, but found no evidence for any organic materials at all.

This was very puzzling. Even the sterile lunar samples returned by the Apollo astronauts contain some organic materials, mainly from meteorites. So there must be something in the Martian soil that destroys organic material. This would make it difficult for any form of life to survive. Perhaps the ultraviolet component of sunlight, which reaches the Martian surface unchecked by any ozone layer, is responsible.

Above: The second Viking lander site. Trenches dug in the soil by the sampler arm can be seen. The discarded protective canister that shielded the sampler head en route *to Mars can be seen at right.*

The Viking sampling head was designed to scoop and trench samples of soil: compare the pre-mission artist's impression (below) with an actual lander photograph (above).

Then it was suggested that Martian microbes could be cannibals that consume the carbon-containing remains of their own kind: but this seems a highly contrived explanation.

Most scientists now agree that the Viking landers failed to detect life in the Martian soil. Yet there are dissenters: Gilbert Levin, head of the LR experiment, announced at a conference to mark the 10th anniversary of the Viking landing that in the intervening decade no chemical reactions had been identified that could explain the labelled release results. But the consensus is that an explanation in terms of life is unlikely.

Left: No signs of life were detected by the cameras of either lander in over two Martian years of operation.

Above: Close-up view of the discarded sampler head canister on the Martian surface.

Biological oases

The Viking results, taken from only two sites, are severely limited. They do not rule out the possibility that life exists elsewhere on the surface or beneath it, or that it has existed in past epochs.

Astronauts on the surface would be well equipped to resolve many unanswered questions. To discover whether organic material exists beneath the surface they could systematically drill core sections and analyse them by exhaustive biochemical experiments.

In harsh terrestrial environments, such as Antarctica, microbes have retreated inside rocks: perhaps Martian microbes did the same. Astronauts could pick rocks up and crack them open to look for traces of life.

Viking's investigation of the inorganic chemistry of the Martian soil was also limited. Its X-ray fluorescence spectrometer was not capable of measuring the amounts of biologically important elements such as hydrogen, oxygen and nitrogen. However, the GCMS revealed that between

The sampler arms of the Viking landers were used to move rocks around in the hope that organic material could be detected underneath them.

two and three per cent of the Martian atmosphere is nitrogen, which raised hopes that the building blocks for life might be present.

If there is life on Mars today, where is it most likely to be found? The most probable answer is: wherever the water is. This would imply that there is a better chance of finding life at the permanent northern polar ice cap, which is believed to be composed mainly of water ice. The first data from these regions will be provided by penetrators fired from orbit and from later rover traverses. Organic materials, or possibly microbe colonies, might be found within specific locations.

What are the chances of finding such self-contained ecological niches elsewhere on Mars? Western expert opinion is divided, particularly as the infrared instruments on board the Viking orbiters should have been easily capable of finding large-scale 'watering holes' or 'hot spots' indicating metabolic activity.

Soviet scientists, however, are confident that such 'oases' are present, and have suggested the large crater Lomonosov, at a latitude of 65° north, as a likely site for investigation. It is appropriate

that this crater is named after a pioneering 18th-century Russian chemist: his latter-day counterparts are confident that even if they find no micro-organisms they may find substances such as amino acids from which self-reproducing systems could have developed.

What if life once existed on Mars but has died out? The remnants of such life may be found as fossils or as telltale 'signatures' in organic material, almost anywhere on the surface, but since we know so little of the chemical and atmospheric evolution of Mars, interpreting such finds would be highly speculative. The sedimentary layers within the Valles Marineris canyons and at the polar caps are likely areas for fossils. Traces of extinct life forms may also be preserved in the floors and walls of ancient water channels.

The question of life on Mars will take many years of investigation to answer. The result will have profound implications. If there is no evidence that life ever evolved on Mars, it will throw into doubt our understanding of how terrestrial life evolved and theories of chemical evolution. But if signs of life are found, it could imply that life is not unique and is common throughout the universe.

The ice fields around the north pole remain the region of Mars in which primordial life forms are most likely to be found.

On foot, explorers will be able to venture little more than one mile (1.6 kilometres) from their base. Even with a rover like that used on the Apollo missions, they will be able to explore only a minute fraction of the 56 million square miles (145 million square kilometres) of the Martian surface. But this area could be hugely increased if several automatic roving vehicles were taken along. These could be dispatched on long trips under the control of the surface base or the mother ship. Some could return to base in due course, carrying samples; others could go further afield on one-way trips.

The Martian atmosphere, although very thin compared to our own, is sufficient to allow aeroplanes and balloons to take to the skies. These will be very useful innovations in exploring the red planet, avoiding the problems of obstacles on the surface.

By a combination of rovers, balloons and aeroplanes, the first crews on Mars will be able to explore beyond the range of direct human contact.

Roving vehicles

Travelling across the Martian surface presents many difficulties for roving vehicles (see box). Even the relatively smooth Viking sites revealed large boulders and dust-covered inclines that would be hazardous to both manned and unmanned rovers. A manned mission will probably carry both types of rover to expand the range of exploration. Unmanned rovers might be based on current Soviet six-wheeled, nuclear-powered designs, weighing several hundred pounds and equipped with a standard set of sensors. Several would be dispatched and controlled from the lander, possibly via an orbiting satellite. After several weeks or

Prototypes for unmanned Mars rovers have already been tested by NASA (seen here) and the Soviets.

months of probing far from the base they would bring back samples. They could continue their travels under Earth command for a year or more after the crew's departure.

A larger manned rover will also be essential. One study has suggested that a vehicle for transporting a 1500-pound (680-kilogram) load on a 25-mile (40-kilometre) round trip should be able to cope with 20° slopes and climb over one-foot (30-centimetre) boulders.

The two astronauts, perched on top of a buggy based on the Apollo Lunar Rover design, would depend on their spacesuit backpacks for life support. These might last a maximum of four hours. So a more advanced vehicle might be used, with an Earth weight of a ton, a range of 78 miles (125 kilometres) and perhaps six wheels. It would have its own life-support system, to which the astronauts could attach their EVA suits. It could even provide a pressurized tent that could be erected to enable the travellers to sleep with their spacesuits off on long trips.

Even rovers of this kind would be limited. When using them, samples and cores would still need to be brought back to base for analysis. For more extensive exploration a self-contained mobile laboratory, or Mobilab, would provide a pressurized 'shirtsleeve' environment for at least two astronauts over several days. It might have an Earth weight of 4–5 tons, and be able to transport two or three crew members for up to 30 days. It would have a range of 60 miles (100 kilometres) and a top speed of 20 miles per hour (32 kilometres per hour), and carry two tons of equipment. The crew would find their position by onboard inertial guidance systems and dead

Surface rovers

Roving vehicles on Mars will have to be extremely rugged and reliable to cope with terrain whose hazards are little understood. Design studies have specified some of the requirements they will have to meet.

NASA studies suggest that rovers should be able to travel half a mile (one kilometre) per day throughout a three-year period. They should be able to climb over boulders five feet (1.5 metres) high, and to negotiate inclines of 35 per cent in dunefields and 60 per cent on firmer ground.

Many ideas for rover designs have emerged. One of the most ingenious was a large mesh ball, which would contain scientific instruments and would be rolled around by the winds. However, it soon became clear that such a device would end up in ravines or gulleys with no means of escape.

But what about providing a rolling-ball rover with its own power? This idea has been taken up by graduate students in the Lunar and Planetary Laboratory at the University of Arizona. Their 'Mars Ball' consists of two wheels, each 16 feet (five metres) across and made of 16 tough gasbags. The wheels are linked by a hub carrying the control system and a large airtight chamber that stores compressed atmospheric gas. The bags are individually inflated and deflated, making the wheel roll.

The Mars Ball has the advantages of simplicity and ease of control, but its reliability and value as a scientific platform have yet to be demonstrated. For over a decade NASA's Jet Propulsion Laboratory has been studying more conventional rover designs, akin to the Soviet Lunokhods and the American Lunar Roving Vehicles.

Preliminary studies indicate that such rovers should have six independently driven wheels and an articulated body, divided into three 'cabs' for mobility over obstacles. The rear cab would

carry a nuclear power supply. A stereoscopic TV camera would protrude from the central section, which would also house the command control systems. On the

The University of Arizona's 'Mars Ball' moves by a controlled cycle of inflation and deflation of its radial gasbags. Its control electronics and scientific equipment are mounted on the hub.

front cab would be mounted a sampling drill and a pair of robot arms to which a variety of tools could be attached. These would be specialized for tasks such as trenching or sifting the soil.

Control of the rover would be a major problem. If it were manually controlled from the Earth, as were the two Soviet Lunokhods, it would have to halt every few feet to check for hazards. The average delay between transmission of a signal from the rover and the arrival of the response from the Earth would be about 25 minutes: it would take all day to advance a hundred yards. That is why manual control from the lander, the orbiter or the Martian moons is such an attractive idea.

A completely autonomous rover that makes its own decisions and automatically avoids dangers is beyond current computing ability. But a semi-autonomous vehicle is feasible. A system known as CARD (Computer-Aided Remote Driving) is being pioneered by the US Army for its advanced vehicles.

A rover using this technology would transmit stereoscopic TV pictures to the Earth, where computers would work out the next 300–800 feet (100–250 metres) of its path, so that it could avoid boulders and crevasses.

A more advanced semi-autonomous technique is also possible. Stereoscopic TV pictures showing surface details as small as three feet (one metre) across would be sent from orbit to the Earth. There, CARD computers would work out a hazard-free route for the rover, which would stop every 30 feet (10 metres) or so to compare what it was seeing with what it expected to see. If it encountered an unexpected hazard it would move around it and return to the planned route as soon as it could safely do so. If the problem was beyond its computing abilities it would stop and wait for further instructions from the Earth. In time CARD computers could be set up at Mars bases, speeding up the exploration of the planet.

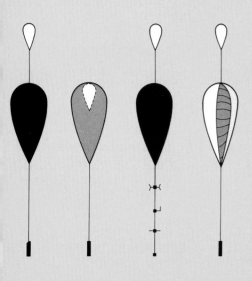

Magnificent flying machines

Aviation on Mars presents many unique problems because its atmosphere is so thin. Balloons displace atmospheric gas to gain buoyancy; aeroplane wing surfaces produce aerodynamic lift by their movement. On Mars each craft has its own advantages and disadvantages.

The atmospheric pressure at the Martian surface varies between five and seven millibars—less than a hundredth of that on Earth. Such low values are found at about 100,000 feet (30 kilometres) above the Earth, where aircraft can hardly sustain flight. Another difficulty arises because the atmosphere is mainly carbon dioxide: no practical fuel can burn in it. So if an aircraft uses engines it will have to carry oxidizer supplies, thereby reducing the amount of payload that can be carried.

The lower Martian gravity is a mixed blessing. For balloons, it decreases the effect of buoyancy. For aircraft, by reducing the weight to be lifted, it effectively reduces the equivalent terrestrial altitude by a third, putting it well within the range of high-altitude planes. The very thinness of the atmosphere also makes it easier to propel a vehicle through it.

Engineers at NASA's Jet Propulsion Laboratory are testing the 'dual balloon' design proposed by Jacques Blamont (see main text) and are pleased with the results so far. But as any balloonist on Earth knows only too well, ballooning presents many problems. Wind conditions are the most worrying; it is difficult to inflate any balloon in windy conditions. For this reason it has been suggested that the helium balloon be placed inside the open balloon, thus holding the latter open and allowing it to fill out more quickly.

The problems of inflation may be circumvented during the early unmanned missions, such as the Soviet missions in the 1990s:

balloons may be dropped from orbit. The larger balloon will unfurl and fill as it falls, with the helium balloon acting as a parachute.

Another worry is that the scientific payload beneath the balloon could become snarled up around surface rocks. This risk can be reduced if, rather than having the instruments located in one self-contained package, they are strung individually along a rope 300 feet (100 metres) long. In fact the instrument rope could act as an excellent transmission aerial, the longer the better.

However, data transmission will be limited by the balloon's power, realistically expected to be only a few watts. This will limit the amount of data that can be returned directly, so communications will have to be routed via an orbiter. The Soviets have said that during their 1992/94 mission, data could be sent at a rate of around 15 kilobits per second for a few hours each day, via the orbiting spacecraft.

The balloon's lifetime will be limited by the inevitable diffusion of helium through the skin of the uppermost balloon. Could other gases be used? Hydrogen is easier to contain, and has the advantage of being carried on the mission in liquid form as rocket fuel.

Controlling height

Blamont's design uses an open-bottomed balloon to allow atmospheric gas to flow in and out, permitting a daily cycle of rising and sinking. (A sealed balloon can rise only to an altitude at which the expansion of the internal gas makes its skin taut.) Control of the balloon's height would be achieved by use of a valve at its top, which would vent atmospheric gas and decrease the balloon's buoyancy. By venting the gas in this manner the balloon could use wind currents at different altitudes to travel in the desired direction: the guide rope would help in directional control by acting as a keel.

Detailed views of the surface would be returned by balloons above Mars. At night they would descend to allow instrument packages to analyse the soil directly.

Mars balloon variants (from left): the basic 'dual balloon'; the helium balloon placed within the solar balloon; instruments strung along the guide-rope: a 'racing' balloon with greater manoeuvrability.

But venting would provide only crude manoeuvrability. It has been suggested that to control the path of a balloon, a design similar to that of manned racing balloons be adopted. One half of the balloon is painted silver on the outside: the remainder is clear. The silver half would be painted black on the inside. Small motors would turn the balloon, so that the silvered side was towards or away from the Sun; this would vary the heating effect and would control the operating altitude.

Marsplanes

Aeroplanes on Mars will have to contend with the lower atmospheric density. A given plane design will have to move at least six times faster on Mars than it would on Earth to generate the same aerodynamic lift. About 2½ times as much power would also be required. An air-breathing engine cannot be used, and battery-powered propulsion would be limited in range, as would unpowered glider designs.

The speed of sound on Mars is only around 450 miles per hour (720 kilometres per hour), so to avoid breaking the sound barrier the plane's wing area will be large to allow it to fly slowly. Increasing the wingspan increases the aerodynamic lift, but creates problems in folding the wings during the journey to Mars.

A runway is hardly likely during the first manned Mars missions, so a Marsplane will have to be launched by catapults or vertical takeoff and landing (VTOL) rockets.

The most sophisticated form of aviation on Mars could be a manned dirigible or 'blimp'.

During 1977 JPL investigated a number of designs for an unmanned Marsplane. It was concluded that a 660-pound (300-kilogram) sailplane powered by a hydrazine-fuelled piston engine would be capable of reaching a cruise speed of 300 feet per second (90 metres per second). Such a vehicle would cover a range of around 2500 miles (4000 kilometres). There are no reasons why a larger manned version could not be built, but this would have to wait until experience was gained with the unmanned planes.

However, a hybrid buoyant/aerodynamic airship or blimp seems more promising. About 15 per cent of its lift would be derived from buoyancy, the rest generated aerodynamically by its shape. Calculations show that a terrestrial blimp modified for Mars (having the same mass and travelling at the same speed) needs only 22 per cent of the power. This could easily be provided by efficient lightweight solar cells. The blimp would be filled with hydrogen, probably manufactured from the soil.

To be able to lift an astronaut with around 110 pounds (50 kilograms) of equipment, the blimp would be around 350 feet (100 metres) long. It could be steered by simple propellers that could rotate for takeoff and landing.

reckoning, corrected at frequent intervals by fixes from orbiters.

The Mobilab will be designed to permit as much work as possible to be done from the protection of the cabin. There may be protruding sleeves into which an astronaut can slip his arms for work on the surface. EVA suits will need to be taken along for safety. An airlock would be valuable—otherwise the whole cabin will need to be depressurized whenever any crew member needs to work outside.

Batteries alone could not power such a large vehicle and its systems: merely drilling for core samples would require 10 kilowatts of power. Fuel cells could generate electrical energy and heat by the combination of hydrogen and oxygen, with water as a by-product. This water could then be broken down by electrolysis to provide the oxygen and hydrogen fuel for more expeditions.

Balloons

Balloons will be carried on the Soviet unmanned missions in the 1990s, and they will become indispensable for exploring Mars when astronauts arrive there. They have the advantage of being simple, cheap and light, and they can cover thousands of miles in a few weeks. After unpacking each one, the crew would test its instruments by plugging it into standard test equipment, attach its thin plastic canopy and inflate it with helium from a pressure bottle, and then release the balloon to drift freely. The height to which the balloon would rise depends on its volume, the weight of the payload, and the atmospheric temperature, which would vary during the day.

The drawback of a conventional sealed balloon is that its height cannot be controlled. Jacques Blamont has proposed a dual balloon system (see page 26): below a sealed helium balloon would be a black canopy, open at the bottom like a terrestrial hot-air balloon. Each morning, as the Sun

rises, the Martian air inside this balloon would warm up. In two hours the canopy would fill out to its full volume of 140,000 cubic feet (4000 cubic metres). When the internal air is 90°F (50°C) warmer than the air outside, the balloon combination would slowly ascend, reaching a maximum altitude of 3.7 miles (6 kilometres).

The operating altitude could be adjusted by means of a valve at the top of the atmospheric balloon under the control of a radar altimeter. More detailed imaging of the surface could be performed while the balloons were at lower altitudes: at 650–1000 feet (200–300 metres), the cameras could detect objects half an inch (one centimetre) across or smaller.

At night the atmospheric balloon would cool and deflate. The upper balloon, containing helium, would provide slightly less lift than needed to support the whole system's weight, and the balloon would sink. From it would dangle a titanium-clad 'snake', 10 feet (3 metres) long and 5 inches (13 centimetres) thick. This would contain batteries, transmitters, heaters and scientific instruments. When the snake touched the ground, the balloon's descent would be checked. While the snake provided a surface anchor, its instruments would make measurements.

The lifetime of such an experiment would depend mainly on battery capacity, and could perhaps stretch up to two months, enabling a balloon to travel half-way around the planet. Their movement could be tracked from the Earth by linking the observations of radio telescopes in different continents, a technique called very-long-baseline interferometry (VLBI). The technique was successfully demonstrated with the Soviet Venus balloons in 1985: speeds could be measured to within two miles per hour (three kilometres per hour), and positions to within half a mile (one kilometre).

The Marsplane

The attractive simplicity of the balloon will be outweighed in later missions by its minimal controllability. Studies have shown that a pilotless powered aircraft could undertake reconnaissance, drop scientific packages and penetrators, and even deliver materials to exploring parties.

The Marsplane would be bolted together by the crew after landing. It would look like a large powered glider, and would be launched by a catapult or by rockets. It would be powered by a 15-horsepower (11.2-kilowatt) piston engine driven by steam generated by the chemical breakdown of hydrazine. The single-bladed propeller would be 15 feet (4.5 metres) from tip to tip. The whole aircraft would have a wingspan of 70 feet (21 metres), a length of 21 feet (6.5 metres) and an Earth weight of 660 pounds (300 kilograms).

While within range of base, the plane would be guided by one of the astronauts; but when out of range, it would be directed by preprogrammed instructions. It could fly as high as five miles (eight kilometres). It could transport a payload of 88 pounds (40 kilograms) for 2500 miles (4000 kilometres), representing 15 hours' flying time. It could land vertically by deliberately stalling, and take off again by means of its onboard hydrazine rocket engines.

The enormous advantages of the Marsplane are its reusability, manoeuvrability and range. Perhaps the day will come, on a later mission, when Mobilabs and manned Marsplanes fan out over the surface of the planet in a concerted programme of scientific exploration.

A typical flight sequence for an unmanned Marsplane. After a catapult- or rocket-assisted launch, the vehicle could cover nearly 2500 miles (4000km) in half a day. On returning to base, it would decelerate by controlled stalling, cushioning its fall by rocket motors.

While the world's attention is focused on the events on Mars, the crew left behind on the main craft must continue with their own work. They will have a multitude of scientific tasks of their own to perform, such as systematic mapping of the surface and close investigation of scientifically interesting areas. There will probably be a separate control room on the Earth specifically for communication with the orbiter.

The departure of the lander will, of course, be a very busy time. Everyone must be prepared for an emergency docking if anything goes drastically wrong. The orbiter will take the lander as low as possible to conserve the smaller vehicle's propellants (unless this job is done by a separate orbital tug). All the time the lander is on the surface the orbiting crew will need to maintain their proficiency in all routines, including possible rescue procedures.

Despite their busy schedule, the orbiter crew will still have to undergo a strenuous exercise regime. Otherwise they could be the unhealthiest members of the expedition on their return, since they will not have had the benefit of a period of Martian gravity.

Communications relay

A vitally important function of the orbiter will be relaying communications between the surface team and the Earth. Direct contact between the surface and mission control will be broken during the Martian night, when the explorers are on the side of the planet turned away from the Earth. This could be especially serious for future missions near the poles during the long nights of winter. However, during each of these night-time periods the orbiter will be in sight of both the explorers and the Earth for much of the time and will therefore be able to relay signals between them.

Nevertheless, there will still be some 'dead' periods. It will be necessary to maintain a complete and continuous communications link between lander, orbiter and Earth. A network of at least three small orbiting communications satellites (comsats) in equatorial stationary orbits would provide this.

Finding the lander

One of the first tasks of the orbiter crew must be to pinpoint the landing site. Geologists will need to know it precisely in order to fit the surface observations and

samples into their surroundings. Furthermore, the ascent stage guidance computer will be able to produce a more accurate return orbit if it knows the exact takeoff point. The radio navigation beams used to 'talk down' the lander will allow accurate pinpointing of its position. In the event of discrepancies in the beacon data, the crew will look for the lander with a low-powered telescope while they are near the low point of their orbit.

Calibrating orbiter data

The results gained by the Mars surface explorers will be essential to the interpretation of the enormous quantity of data that will be collected by remote sensing instruments on the orbiter. Spectrometers will analyse the visible and infrared light reflected from the surface to determine the surface minerals. A gamma-ray spectrometer will search for the presence of certain radioactive elements, such as uranium. Radar can measure the height profile of the surface, and can also probe beneath the surface layers.

High-resolution pictures of the landing site will help mission control to plan extravehicular activity (EVA). The scans will preferably be performed at the orbital low point for the greatest resolution. The path of the orbiter on this first manned mission is likely to lie close to the equator, but on later missions a high-inclination orbit might be possible and would permit more of the surface to be scanned.

One important task of the orbiter crew will be to study the planet's weather patterns, particularly the development of localized dust storms, which may

Photographing Mars from orbit will be an important task for the orbiter's crew. Large film cameras, like this one, will permit far higher resolution than TV pictures.

Above: Communications between the lander and the orbiter would be highly efficient if a network of communications satellites, like the one shown, were deployed. Scientific satellites could also be released.

Occasional spacewalks by the orbiter astronauts will be necessary for minor repairs and the maintenance of scientific satellites.

Releasing communications or other satellites could be performed manually by astronauts.

interfere with surface activities.

The orbiter crew will also be free to photograph areas that catch their interest, just as the Apollo pilots did. The lunar programme demonstrated that such pictures can be vital in picking out sites for future investigation.

Many other experiments will call on the crew's time. Several subsatellites could be released into low orbit to investigate charged particles around the planet by measuring their effects on radio waves beamed to the parent craft from the satellites. The orbiter might also collect data from a penetrator fired into the surface in the polar regions, where a first expedition will not venture. The atmosphere and the planet's weak magnetic field will come under scrutiny from a battery of instruments, and micrometeoroid detectors will assess the dust surrounding the planet.

Astronomy from orbit

The work of the crew left in orbit may not seem as glamorous as that of their colleagues on the surface. But there are many areas of investigation that can be opened up, such as extending astronomy far into the solar system. For example, a 50-foot (15-metre) radio telescope, operating in conjunction with Earth-based dishes, will 'see' with detail that is hundreds of thousands of times greater than is presently possible.

The only limit to research capabilities will be the amount of payload that can be accommodated on the orbiter. There will be many teams of scientists clamouring to have their experiments placed on the orbiter, but only a few can be accommodated. Whatever selection of experiments is finally chosen, it is certain that the orbiter crew will be far from idle while their comrades are exploring the surface of the planet below.

Mission control on Earth will constantly monitor the crew activities, but the lengthy signal delay will prevent casual conversation.

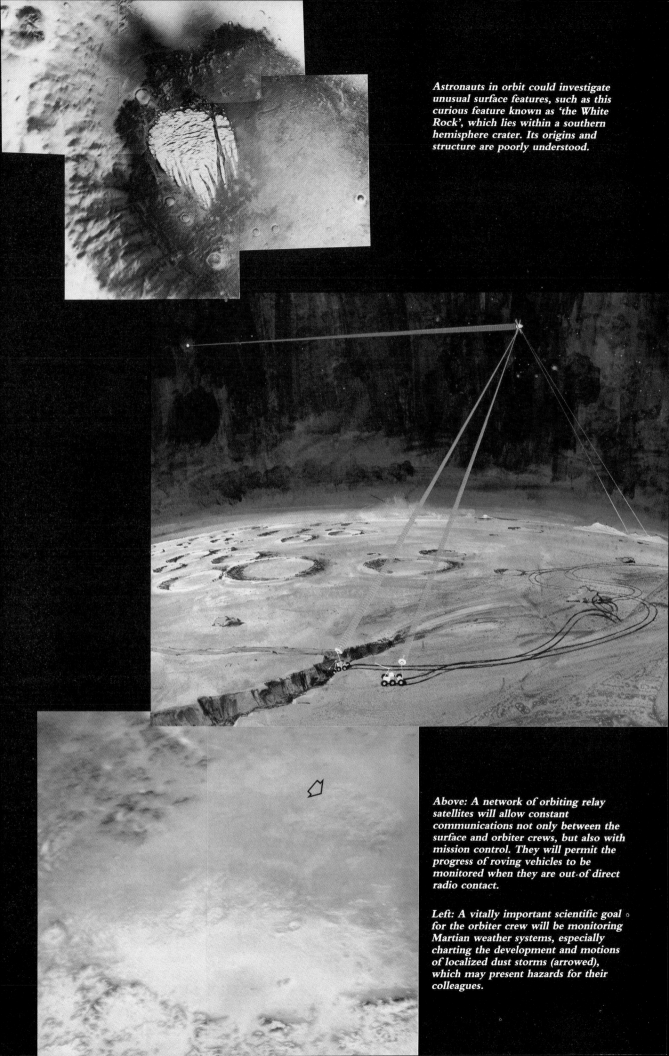

Astronauts in orbit could investigate unusual surface features, such as this curious feature known as 'the White Rock', which lies within a southern hemisphere crater. Its origins and structure are poorly understood.

Above: *A network of orbiting relay satellites will allow constant communications not only between the surface and orbiter crews, but also with mission control. They will permit the progress of roving vehicles to be monitored when they are out-of direct radio contact.*

Left: *A vitally important scientific goal for the orbiter crew will be monitoring Martian weather systems, especially charting the development and motions of localized dust storms (arrowed), which may present hazards for their colleagues.*

At least one early manned mission to Mars will include the planet's two small moons in its itinerary. Apart from the purely scientific interest of examining the bodies themselves, they will probably provide important bases in their own right. Eventually they may even supply oxygen and propellants for trips to the surface and the journey home.

Rendezvous with the moons

The orbit of Deimos is 12,470 miles (20,100 kilometres) high; that of Phobos is 3720 miles (6000 kilometres) high. The spacecraft will rendezvous with them in the same way as the Soviet Phobos craft (see pages 18–25). On arriving at Mars it will go into an elliptical parking orbit. Then, igniting its engines near the high point of this orbit, it will raise the low point to the satellite's altitude. At the same time it will change its orbital tilt to match that of the satellite. At the new low point it will fire its engines again to lower its high point, creating a circular rendezvous orbit that is almost identical to the satellite's.

It would require a $\triangle v$ (see page 78) of about 2100 feet per second (650 metres per second) to reach Deimos from a parking orbit with a 310-mile (500-kilometre) low point; and of 2100–2800 feet per second (650–850 metres per second) to reach Phobos. The precise figures would depend on the shape and tilt of the initial parking orbit. If the latter had a different tilt from the satellites' orbits, the total $\triangle v$ required could be pushed up to 0.6 miles per second (one kilometre per second). This is a significant amount, but on the other hand less fuel would be needed for the return to Earth than from the original parking orbit.

Mars, its surface scarred by the great equatorial volcanoes and canyons, looms over Phobos in this artist's impression.

The Soviet Phobos craft will return vital information about the raw materials that the satellite has to offer.

Once the spacecraft is in an orbit that closely matches the satellite, there will be a period of optical observation to pin down the satellite's position precisely. Then the final approach will begin. The astronauts will probably pull their craft up short and fly around the moon a few times to bring their batteries of remote sensing equipment to bear. The manned flyaround permits an observing period much longer than that of the unmanned Phobos mission, with larger and more sensitive instruments, so the images will have much greater coverage and resolution.

Harpooning a moon

The crew will fire harpoons into the surface to act as anchors. The approach will need to be very gentle, not only to avoid damage to the lander but to avoid the danger of rebounding. Each moon's rotation has been 'captured' by the gravity of Mars so that one end of the elongated body always points towards the planet. It is probably this 'lower'

end that the lander will touch down on.

From Phobos, Mars will look huge, about 80 times larger than the Moon looks from the Earth. It will bathe the satellite in its red glow during the Phobos night. The astronauts will see Mars change from full to crescent and back to full every 7.6 hours. Seen from Deimos Mars will look about 30 times the size of the Moon in our skies, and will go through its cycle of phases in just over 32 hours.

Walking on Phobos will be an interesting experience, since its surface gravity is about a thousandth of the Earth's and the strength of its surface soil is very low. Apollo astronauts had to contend with lunar gravity one sixth of Earth's. They found the best way forward was in a series of kangaroolike hops. They frequently lost their balance when turning or trying to stop. The problems on Phobos will be much greater. The walkers will have to go extremely slowly or they will bounce upwards and have to wait a few seconds to come down. One advantage is that they will be able to carry huge packs around. An object weighing 500 pounds (227 kilograms) on the Earth weighs only half a pound (227 grams) on Phobos. The escape speed is about 13 miles per hour (21 kilometres per hour) so an astronaut could jump several hundred feet high, requiring several minutes to drift downwards again. The crew could certainly throw objects into space. Such feats would be somewhat easier on the slightly less massive Deimos.

Propellant manufacture

The Soviet Phobos mission will return much data about the structure and composition of the larger moon. If, as now seems likely, they are C-type asteroids, they may contain about 20 per cent by weight of water, and some carbon. This raises the possibility of on-the-spot manufacture of oxygen, water and propellants. It has been estimated that an oxygen-producing plant would have an Earth weight of 50 tons, which would be extended by 25 tons if it were to make fuels as well. These are large masses to carry from the Earth, but in the long term the investment would pay off, since spacecraft could travel to Mars from the Earth without carrying oxygen or propellants for the landing or the return journey. The mass of a Mars craft starting its voyage from Earth orbit could be cut by a third.

Alternatively, Phobos could provide supplies for a base on the Earth's Moon. It would require only one third as much fuel to ship materials from Phobos to the Moon as to carry them there from Earth, battling against our planet's relatively strong gravity. The drawback is the unavoidably long transfer time from Phobos.

Bridgehead on Deimos

It has been suggested that it would be more sensible to explore the two moons of Mars before attempting to land on the planet. It is certainly easier to brake into the high orbit of Deimos than into a parking orbit prior to a landing on Mars. Once propellant manufacture had been established, a lander could first go to Deimos, manufacture propellants from its raw materials and then make a burn with a $\triangle v$ of 1500 miles per hour (2400 kilometres per hour) to descend into the atmosphere and aerobrake into a low circular orbit in preparation for landing.

In fact Deimos would make a good base from which to explore Mars with remotely controlled rovers, sample return craft and penetrators. Habitats could be buried on the satellite for protection from solar flares and cosmic radiation. Deimos moves around the planet in slightly more than one Martian day, which means that any point on Mars is in view for up to 40 hours at a time.

An exotic technique could be employed to save propellant when spacecraft are launched from a satellite. Strong tethers could be extended for hundreds of miles above and below the moon. A craft could haul itself down one of these by an onboard electric motor, with no expenditure of propellant, and then could be released. To prevent disturbances of the system, a vehicle of equivalent mass would have to travel up one of the outward-going tethers at the same time, to be released into a higher orbit.

The Soviets have hinted that they might prefer to establish bases on Phobos and Deimos before attempting to land on the planet. These two 'natural space stations' may prove indispensable in the conquest of Mars.

The exploration of the Martian moons will present scenes reminiscent of the Apollo missions.

THE RETURN

Blast-off from Mars will scatter dust and debris, as in this picture of the ascent of Apollo 14 from the surface of the Moon.

As the time to return approaches, the crew will make the last forays outside the ship. They may store geological equipment on the surface for use by later expeditions. They will make a final check of the scientific instruments that will be left behind—seismometers and meteorological stations, for example.

If the landing site is close to one of the Viking craft, the astronauts may take pieces of it back with them, so that engineers on Earth can assess how well it has stood up to Martian conditions since 1976. They may switch the rovers to remote-control mode, so that they can continue their surveys. But the rovers' first job may well be to record, from a safe distance, the first manned launch from Mars.

Rubbish left behind must be carefully bagged to avoid contaminating the planet. All the surface samples that are being taken back will have been accurately weighed to ensure that the ascent vehicle is not overloaded. Unwanted overboots, backpacks and so on will be jettisoned to dispose of all unnecessary weight for the ascent to orbit.

The takeoff

The ascent stage systems will be fully powered up, at first drawing on the electrical supply from the lower stage. Just before launch they will switch to the ascent stage's batteries. The computers will begin their launch routine at the appointed time, pressurizing the propellant tanks and actuating valves. But the sequence will be overseen by the crew at every step.

At the last moment, propellants will be forced through into the combustion chambers and the ascent stage will sever all connections with the lower stage by 'pyrotechnic guillotines'—explosive charges that slice through electrical and hydraulic lines and mechanical attachment points. The ascent stage will rise vertically until it has passed through the denser lower layers of the atmosphere, and then tilt to head for the chosen orbit.

The chance of failure

If the ascent stage engines do not fire, the crew will be stranded, and the orbiter crew and Earth control will be powerless to help them. To preclude this, the design of the engines will be kept as simple as possible. Propellants such as nitrogen tetroxide and hydrazine will be used, because they burn on contact, with no need for an igniter.

If the engines *do* fail, the crew will of course set about trying to repair them. They will have several days before the orbit of the mother ship (or orbital tug) is correctly positioned, directly overhead, for a second attempt.

A computer failure *during* takeoff will be somewhat less serious than during landing, since it is easier to get into an emergency orbit than it is to land safely. In this event the tug or orbiter would have to come down to the lander's orbit for the rendezvous.

Entering orbit

The lander must reach a speed of about 10,770 miles per hour (17,300 kilometres per hour) to enter low Mars orbit, with a high point perhaps 120–310 miles (200–500 kilometres) above the surface. Once in orbit, the ascent stage will fire its engines again to lift the low point well clear of the atmosphere.

The reunion with the orbiting craft might be accomplished in a variety of ways. The lander might join a tug in this first low orbit: in that case the tug would then fire its own engines for rendezvous with the orbiter. Or, if a tug is not used, the main craft might drop to a lower orbit to meet the ascent stage.

Once the craft have mated and airtightness has been verified, the hatches will be opened and the two crews can greet each other after their months of separation.

Unloading the lander's cargo and putting the samples in their prepared places on board the orbiter is likely to be a time-consuming job. Waste and items not required for the return will be stored in the lander, which will be placed in a parking orbit. But spacesuits will be retained for the homeward journey, in case outside repairs should be necessary. The tug, if there is one, will also be left in a parking orbit, since it could be useful to later missions.

The launch towards Earth

There is some leeway in the timing of the trans-Earth injection burn, depending on the propellant available: the launch window might be 10–30 days long. But it is crucial that this opportunity should be taken: if it were to be missed, there would not be another for about two years.

The departure from Mars parking orbit will be almost the reverse of arrival. Several minutes' firing will free the spacecraft from the grip of Mars and put it into an orbit around the Sun—but its speed will be lower than that of Mars, so that it will begin to fall inwards towards the Sun. The required Δv will be about 4000 miles per hour (6500 kilometres per hour), depending on the height of the parking orbit from which the craft departs. Once the engines have closed down, the computer will issue an orbit prediction and, if all has gone well, declare the spacecraft is on an Earthbound trajectory.

The lander ascent stage separates to head off into rendezvous with the orbiter craft.

The familiar sight of their home planet will greet the astronauts after an absence that may be as long as three years.

Quarantine

The months-long return flight will be much like the outward leg, except that much of the work will be concerned with preliminary analysis of the newly gathered data from Mars. But now there will be an additional reason to keep a close watch on the crew's health. If there are harmful micro-organisms on Mars, the explorers will almost certainly have been contaminated, and disease will appear during the journey home. As a check, mice may be carried on the mission as test subjects: on the journey home they would be exposed to soil samples and then kept under observation.

Conceivably, samples returned by unmanned missions might already have demonstrated conclusively that there is no danger to the Earth from Martian materials. But it is more likely that returning explorers will have to undergo weeks of quarantine. NASA's plans for the arrival at Earth are as follows.

Above: A piece of Mars? The similarities between this meteorite sample and what we know of the Martian rocks suggest that it may have originated on the planet. It should be possible to settle this question when the first Martian rock samples are analysed. Earth recovery operations will be conducted on a space station (below), with a separate quarantine facility.

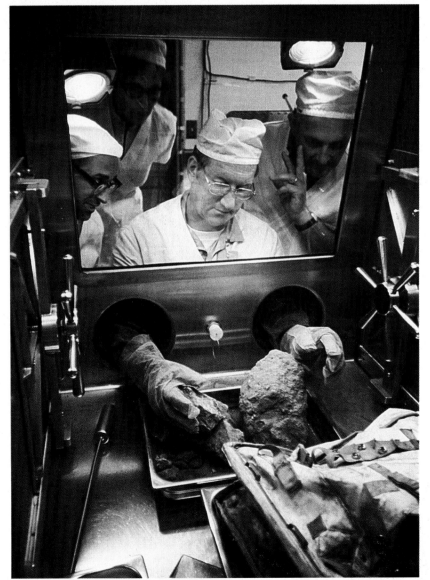

After the quarantine procedure, the first Mars explorers will return to the Earth for a full debriefing and a heroes' welcome.

The returning spacecraft will be guided into Earth's upper atmosphere, plunging in at 29,000 miles per hour (45,000 kilometres per hour). Aerobraking will cut the speed by more than a third, though peak deceleration will be held to 5g or less, because of the stress on astronauts who have experienced months of weightlessness.

When the craft has been slowed sufficiently and is in the right orbit, it will dock with an orbital transfer vehicle, a tug that will propel it to the Orbital Quarantine Facility (OQF), stationed in low Earth orbit.

If the crew is found to have been contaminated, and the threat to the Earth is thought great enough, they might have to remain aboard the Mars craft for another two months while Mars samples are examined. But more probably they will need only to swab down with disinfectant towels before donning biological protection garments and transferring to sterile quarters on the OQF. They will be monitored by doctors while they pass their time with extensive debriefings, family linkups, interviews and recreations. When the testing of the Mars samples has been completed and they have been declared contamination-free, the crew can at long last return to Earth and a triumphal reception.

Mars samples will be diligently stored to avoid any contamination during quarantine and subsequent analysis.

A testing time

The tests on the Mars samples will be carried out in the Orbital Quarantine Facility's laboratory module. This is an isolation unit where a team of specialists can study a few ounces of Mars samples for biologically active agents. The whole module is kept at a slightly lower pressure than the rest of the OQF so that any air leakage will be inwards. All waste products from the lab module's activities will be sealed and stored. Operators will go through decontamination procedures on entering and leaving the module. Ultraviolet light will bathe the decontamination chamber when it is unoccupied. If the lab should become seriously contaminated, astronauts would don protective suits to clean up, flooding the module with lethal formaldehyde gas.

The module's central aisle will be lined with sealed cabinets, containing arrays of instruments for simulating the thin, cold carbon dioxide atmosphere of Mars.

Small portions of Martian material will be removed from the return craft in sealed containers and passed into the lab module through a sterile airlock. Inside the cabinets samples will be handled with mechanical manipulators. A stream of air will flow down the front of each cabinet to carry any escaping material away from the operator, who will be wearing masks and gowns.

A canister containing a sample will be placed inside a cabinet. The exterior may have been contaminated by Earth organisms, so the outer layer may be burned off. A piece will be ground down into particles less than a two-thousandth of an inch (a hundredth of a millimetre) in diameter and studied under microscopes for traces of fossil life.

Selected samples will now be put through a sequence of five tests for life:

In *phase 1* measurements will be made of various quantities that will be important to later tests, such as acidity/alkalinity. There will also be tests for amino acids, the building blocks of life on the Earth.

In *phase 2* workers will search through microscopes for cells adhering to soil particles. If any are found, they will be stained to highlight specific cellular features.

In *phase 3* soil samples from different depths will be dosed with a range of nutrients in a variety of environmental conditions to see whether changes characteristic of life occur.

In *phase 4* attempts will be made to grow Mars organisms by incubating samples on culture plates with likely food material.

Phase 5 is the final, critical 'challenge test', in which Mars organisms will be given the opportunity to attack terrestrial biological materials, such as lung and kidney tissue, over about three weeks.

If no activity is detected, all material can be passed for study by Earth laboratories, and the crew can at long last go home. If there is a response, the action taken will depend on its precise nature. But if the Mars crew has returned in a healthy state, no one will be expecting any serious hazard from the samples.

Below: The NASA Space Station modules such as those designed by Boeing will provide the basis for an Orbiting Quarantine Facility.

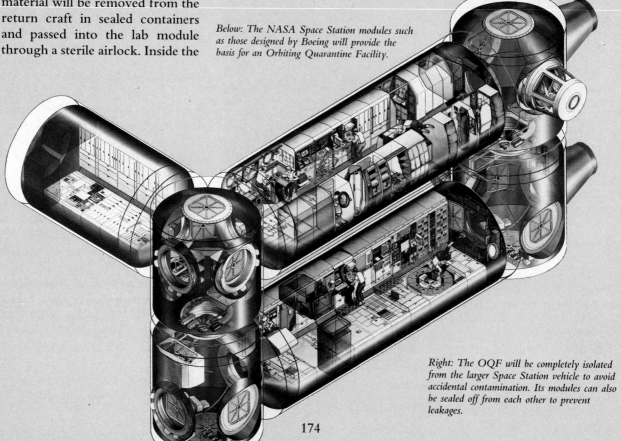

Right: The OQF will be completely isolated from the larger Space Station vehicle to avoid accidental contamination. Its modules can also be sealed off from each other to prevent leakages.

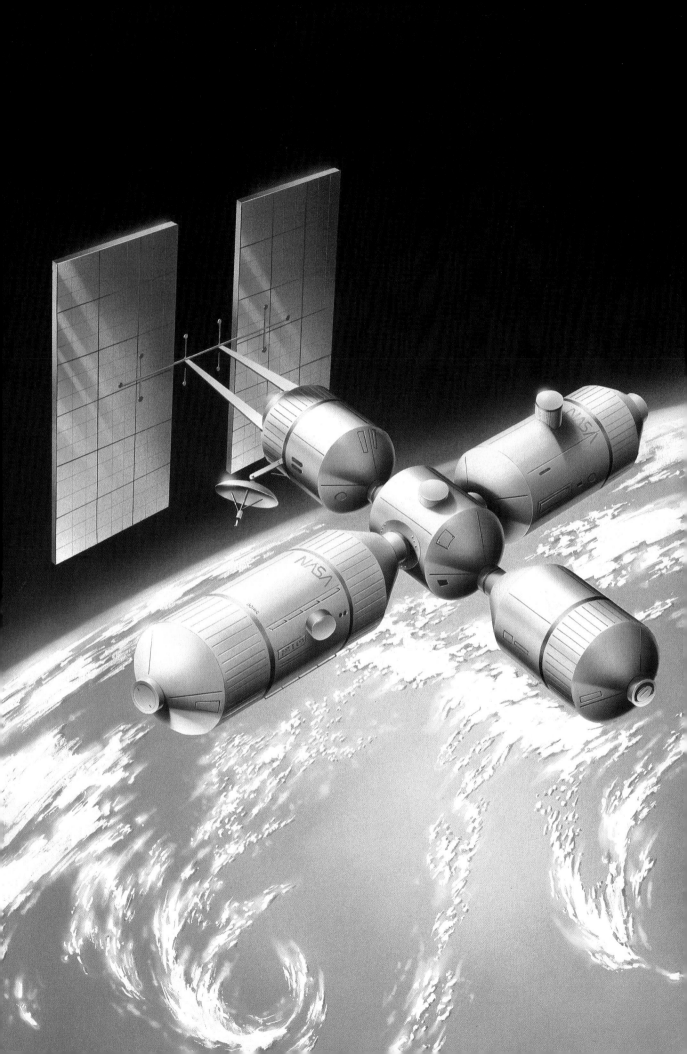

PART SIX
THE FUTURE

A PERMANENT BASE

A permanent base on Mars will undoubtedly be set up one day. In 1986, the US National Commission on Space recommended that a full base with a crew of 20 should be established by the year 2030, 15 years after the first outpost. But Mars will be thoroughly explored before planners plump for a permanent site: they do not wish to invest huge sums in one place only to find a few years later that there is a far better location not far away. They might conclude that Phobos and Deimos should be developed first: some kind of orbiting station will probably be required to support a surface base and the moons could fill this role—especially if they should prove to be rich in raw materials.

It would be convenient if the first landing site were suitable to be the nucleus of a permanent base. But this is not likely: it will be chosen primarily on grounds of safety and the immediate needs of scientific research. Some experts have suggested that three landings, at different sites, will be necessary before a base site is selected. The missions would be sent not in successive launch windows but at four-year intervals, so that the later missions would leave after the previous crews had returned and their data had been studied in detail.

Each site would be thoroughly

In situ propellant production

When the first Mars explorers depart from the surface, their engines will burn liquid propellants transported at great cost all the way from the Earth. Until they are called on, these materials and the tanks that hold them are just so much deadweight. It would clearly be far more efficient to obtain the propellants on Mars itself. The mass of the spacecraft on departure from Earth orbit might then be halved.

In Situ Propellant Production (ISPP) would need to make use of readily available and easily handled raw materials. Standard propellants such as nitrogen tetroxide and hydrazine are not suitable for ISPP on Mars because they are made from nitrogen, which is in very short supply. Oxygen production is likely to be the first step: oxgyen is an excellent propellant and is available in large quantities from the carbon dioxide in the Martian atmosphere.

In one such ISPP scheme, heat and power are provided by nuclear generators. Martian air is pumped in, cleaned and compressed about 150-fold to Earth atmospheric pressure, raising its temperature to 660°F (350°C). It then passes to an 'oxygen cell', where it is heated to 1800°F (1000°C). This breaks it down into carbon monoxide and

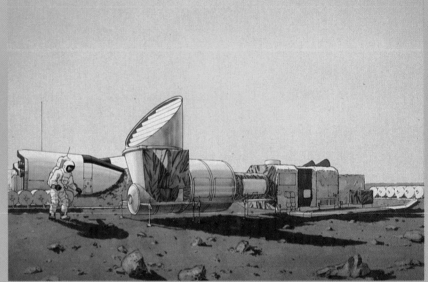

A vital component of a permanent base will be a gas extraction system to convert carbon dioxide into oxygen for life support systems and fuel production (above). The landers would be refuelled in case an emergency return was needed.

oxygen. The latter is compressed to 28 Earth atmospheres and refrigerated at −297°F (−183°C) in a cryogenic tank. (The carbon monoxide might be used as a fuel—though a rather poor one.)

A small unit would have an Earth weight of 660 pounds (300 kilograms), would draw three kilowatts of power and could generate 22 pounds (10 kilograms) of liquid oxygen (lox) daily. It could be placed on the surface by an unmanned sample return craft. For its ascent the craft would be fuelled by onboard stores of liquid methane, burning with the oxygen

provided by the ISPP unit. By the time of takeoff, the unit would have produced enough oxygen to get the craft into orbit.

If the system were scaled up for a manned mission, the unit would have an Earth weight of about 1.6 tons and require 30 kilowatts. It could produce 10 tons of lox in 100 days, which would be sufficient for the ascent stage to return to orbit. But this period would be longer than the 60–80 days stopover of an 'opposition class' mission.

If a source of water is found then ISPP is even more attractive. The atmosphere has so little water that explored, and *in situ* propellant production (ISPP—see box) would be tried out at each. The results could lead to conflicts: for example, scientists might push hard for a visit to some high region, such as Tharsis or even Olympus Mons, but ISPP of oxygen from the atmosphere would be severely hindered by its thinness at these altitudes.

The availability of water could well dictate the base location. Without water it will be more expensive to produce oxygen, drinking water and propellants. Before the landings, permafrost will be sought by remote sensing. Drilling will confirm the depth, extent and accessibility of the frozen water. But it might be discovered that the only useful water sources are around the polar caps: this would be unfortunate, since sites within about 30° of the equator are more easily accessible from orbit.

The earliest missions are likely to be 'conjunction class', with stop-overs of only 60 days (see page 67). For stays of this length, astronauts could live in the lander. But later missions are likely to be 'opposition class', involving a stay of at least a year on Mars, and these will require larger, separate habitation modules. Units already being built for NASA's Space Station by Boeing Aerospace could be used.

vast amounts of air would have to be processed to extract supplies for life-support systems alone. But if a rich water supply such as permafrost is found, then fuel can also be manufactured on the spot, since water can be broken down into hydrogen and oxygen. Although hydrogen can be used directly as a high-energy fuel, it is difficult to liquefy and store. So methane, made by combining the hydrogen with carbon from atmospheric carbon dioxide, is the most attractive candidate.

The advantage of exploiting atmospheric resources is their uniform global availability. It is far simpler to switch on an atmospheric unit whenever and wherever required than it is to search for water and dig it out before beginning conversion to useful products.

On later missions more complex and energy-consuming processes for extracting materials from rocks and soil could be tested. And extraction plants on carbon-rich Phobos and Deimos may be able to produce propellants for the descent to Mars and for the return journey to Earth.

These aluminium cylinders will provide life-support systems, crew quarters and working areas, and will last at least 20 years.

Design for a Mars base

The structures making up the first long-term Mars base are likely to include the following:

a *habitation module* to provide eating, sleeping and leisure facilities in pleasant surroundings;

one or more *laboratory modules* for scientific research;

medical facilities, so that a doctor-astronaut can carry out surgery if needed—it is only a question of time before a major medical emergency strikes a space mission;

a separate *resources module* for life-support units, communications, power and air-conditioning equipment and spares;

greenhouses to contribute food and some oxygen to the base.

Cargo-carrying landers will be employed to build up the first permanent base (above): once emptied, they will become habitat modules in which the settlers will live and work.

As the settlement grows, food will be grown in 'inflatable' greenhouses (left) for self-sufficiency. As a protection against harmful solar radiation, the habitat modules will be buried under the Martian soil (below).

CELSS

The cost of supplying a permanent Mars base with food from the Earth would be enormous. It would be more economical to grow the food where it is needed. To do this a 'Controlled-Environment Life-Support System', or CELSS, would be required.

On long-duration space missions in the near future, oxygen will be obtained from exhaled carbon dioxide and drinking water will be obtained from urine and exhaled water vapour. But in a CELSS, waste materials will be reprocessed by microscopic algae and larger plants, producing food as well as oxygen and water. This would be a tiny version of Earth's ecosystem, a closed system in which plant and human cycles of nutrition and excretion are linked. In theory nothing will be lost and nothing will need to be supplied, though in practice some leakage will be inevitable.

A CELSS will provide a balanced daily diet for each crew member: 3000 calories, 4½ pounds (two kilograms) of water, 2½ ounces (70 grams) of protein and a pound (half a kilogram) of carbohydrates and fats. This can be supplied from an area of about 540 square feet (50 square metres). As on the Earth, the plants will take in carbon dioxide and use the energy of artificial or natural light to make food and produce oxygen.

Some day animals for food might be included in the system, but initially it will be more efficient to have a vegetarian diet of wheat, rice, potatoes, soybeans, peanuts etc. The crops' roots will be supplied with water containing oxygen and dissolved nutrients, while their stems will be supplied with waste carbon dioxide. The portions of the mature plants that are normally considered inedible will be converted into sugar and high-quality protein by microorganisms such as yeast. The remainder can be mixed with human body wastes and burned to

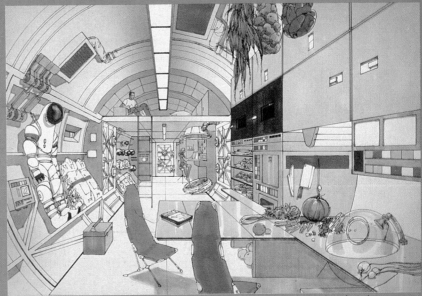

The interior of the first habitat modules: the eventual aim will be for a 'closed' ecological system with the introduction of plants and algae.

generate carbon dioxide.

Most of the elements necessary for plant growth, except phosphorus, have been found in the Martian soil. CELSS is likely to be perfected for the space stations and lunar operations, but on Mars it will have the advantage of the planet's abundant atmospheric carbon dioxide.

The Soviets have long experience of growing plants on their space stations. They have also tested a basic CELSS on the ground. In 1984 two Soviet scientists emerged from a sealed unit called Bios, where they had spent five months growing wheat, peas, dill and other crops under xenon lamps. These plants, grown with nutrients dissolved in water rather

than soil, provided 80 per cent of their food needs. The growing area of 650 square feet (60 square metres) produced more than enough oxygen for them to breathe.

A private American group is building Biosphere 2 (the Earth is Biosphere 1), which covers 2½ acres (one hectare) near Tucson, Arizona. It comprises domes housing areas of jungle, desert, ocean and other environments, and is expected to have a life of 100 years. Up to 10 people at a time will spend two-year periods in the complex. All wastes will be recycled, and electrical power will be generated from solar arrays. Biosphere 2 may provide a preview of a possible future Mars colony.

Soviet cosmonauts have experimented with growing plants in space stations for many years.

All of these structures should be buried to shield them from cosmic rays and solar flares. The base will be equipped with a long-range rover, with a cabin providing shirtsleeve working conditions, for exploration of other sites. One of its first jobs could be as an earth-mover to dig trenches for the modules.

Eventually it might be possible to construct buildings from concrete by using Martian soil. Encouraging results have already been obtained in tests with lunar soil.

Resources for the base

What demands will the base make on Martian resources? If there turn out to be no other suitable sources, water could be extracted from the atmosphere by compressing and refrigerating it. But the process requires 70 kilowatt-hours of energy for each 2¼ pounds (one kilogram) extracted. Permafrost may be a better source. Digging it out would be costly in energy: melting the permafrost with heated underground pipes and then pumping out the water might be the best approach.

At first a base will require about 25 kilowatts of electricity. (The Space Station, by comparison, will initially require 75 kilowatts.) As the base grows its energy requirement will rise to several hundred kilowatts. The Martian winds, though fast-moving, are too feeble to drive wind generators because of the thinness of the air. Fuel cells will be useful for roving vehicles, but their oxygen and hydrogen reactants will be costly, whether they are manufactured by ISPP or brought from the Earth. So fuel cells will not be an economic energy supply for the base as a whole.

A two-ton nuclear generator could be partly buried in a shallow hole well away from the base. Two or even three units would be needed to guarantee reliability. There would be problems of waste heat disposal, as well as the obvious dangers of contamination.

Solar cells would be more efficient on Mars than on the Earth, because they work better at lower temperatures. But they would have their share of problems. Sunlight is weaker on Mars than on the Earth. Radiation damage would cut output by as much as a fifth after 20 years. Dust storms would cut off the sunlight for about 60 days in a Martian year (two Earth years). Solar cells could operate only when the Sun was 15° or more above the horizon: storage batteries must be called on at other times.

It is estimated that for all these reasons a base at 30° latitude would need 1050 square yards (900 square metres) of solar panels to generate the 25 kilowatts average energy requirement—and more to provide a surplus for use at night and so on. Nevertheless, solar cells are an option that is favoured at present.

The base personnel

A staff of at least 12, of both sexes, can be envisaged. They would include scientists, engineers and two pilots (in case one became incapacitated). Two six-person teams working staggered eight-hour shifts is a possible workng arrangement, although it might be desirable to arrange a common sleeping time. Crews will probably be partly changed with each launch window, every two years.

If the first permanent base proves to have been well chosen, it may become the focus of the final phase: the colonization of Mars.

Water reserves on Mars will be a critical factor in establishing a permanent settlement. This Viking 2 lander view shows the sparse frost covering seen in wintertime accounts for negligible amounts of water. The presence of subsurface permafrost and the difficulty in mining it will dictate the choice of the settlement's location.

By the middle of the 21st century, NASA envisages a settlement boasting a full complement of greenhouse, habitat and laboratory modules, with aeroplanes and roving vehicles for advanced exploration.

COLONIZING MARS

If no insurmountable problems emerge during the first tentative years of settlement, Mars will be progressively colonized during the 21st century. Human and robot prospectors will establish distant outposts, making inventories of Martian resources as they do so. Mining, manufacturing and chemical production will follow. Agriculture will develop under pressurized domes to the point where food will no longer need to be imported from the Earth. By the end of the century Mars may well be an independent planet.

Establishing colonies

In February 1988 President Reagan accepted his space policy advisers' plan to develop the technology needed for a return to the Moon or to explore Mars—or both. If Congress approves, NASA will be given $100 million in the financial year beginning in 1989. Propulsion systems, robotics, cryogenic storage methods and aerobraking tests will need to be developed in earnest. Many people feel that NASA should concentrate on the Moon first, developing and improving colonizing techniques. A lunar colony would create a supply base for the more distant planet. In keeping with the Ride report (see page 50), development of technology suitable for both Mars and Moon bases will allow a great deal of flexibility.

A question of transport

The difficulties of transportation between Earth and Mars will limit the initial colonization. Liquid oxygen for cryogenic propellants produced by a lunar base could be supplemented by propellants manufactured on Phobos and Deimos. The two moons will play a key role as transportation and refuelling depots.

Personnel for a long-term base could be ferried to Mars by 'cycling' space stations travelling in VISIT orbits (see page 46). Many of these pioneers will never return to the planet of their birth.

The costs of transporting people, equipment and supplies could be somewhat reduced by using large cargo ships, plying slower routes to Mars, propelled by the pressure of sunlight on vast arrays of sails.

This initial supply phase will require heavy investment to bring the Martian colony to the point where it can build its own robots and factories. The ultimate aim would be to develop robots that could replicate themselves, using raw materials found on Mars or its satellites.

Nuclear rocket engines, using first fission and later fusion, would eventually decrease flight times. However, the solar system will be opened up when antimatter propulsion is developed, enabling spacecraft to cross it in a matter of weeks rather than decades.

Energy from antimatter

Every type of fundamental particle has its 'mirror' particle, or

The Orbital Transfer Vehicle will be crucial for the development of both lunar and Mars bases.

A lunar base will provide vitally important resources for the colonization of the more distant red planet.

antiparticle. Thus protons, which are constituents of the atomic nucleus, have counterparts called antiprotons, which are their opposites in various ways: for example, they are negatively charged, whereas protons are positively charged. Electrons have antiparticles called positrons, and so on.

Antimatter is produced in high-energy particle accelerators but exists only fleetingly: when it comes into direct contact with matter, both are annihilated in a ferocious explosion in which the combined mass is totally converted into energy. The energy released is 1,000 times as efficient as atomic fusion, the process that powers a hydrogen bomb explosion. A thousandth of a gram of hydrogen annihilating an equal quantity of antihydrogen would create an explosion equivalent to 50 tons of TNT. One advantage of the method is that no 'dirty' radioactive debris is produced.

In recent years it has become possible to keep antiparticles circulating in 'storage rings' for many hours. It is only a matter of time before such technology could be employed in specially designed rocket engines. Scientists at the Los Alamos National Laboratory in New Mexico (the institute that oversaw development of the first atomic bomb) reported to NASA in 1986 that antiproton production rates are increasing by a factor of 10 every 2½ years. They believe that a gram of such antiparticles will be produced by 2010—yet if only a milligram could be stored electromagnetically, antiproton propulsion would indeed be feasible.

The routine exploitation of anti-matter as an energy source would have far-reaching consequences for the human race. It would solve many of the world's energy supply problems at a stroke, and would allow spacecraft to travel to or from Mars in a matter of days.

Paradise gained
After its first 30 years, a Martian colony could have a population numbering more than 10,000 and as many as 100,000 by the end of the 21st century. Most of the colonists, and certainly all those who were born on Mars, will have adapted to the planet's weak surface gravity, and would find a visit to the Earth physically distressing. They will know no other life than the domes and subterranean dwellings of their Martian home.

The necessity of wearing spacesuits or being confined within pressurized domes will constrain the development of the Mars colony. Will human beings ever be able to walk unprotected over the surface of the red planet?

There are visionaries who believe that Mars could be 'terra-formed' to produce Earthlike conditions there. If such a process is possible, it would not occur overnight: a conservative estimate

185

Solar sail technology, shown here investigating a comet, will allow economic and efficient transport of cargo to Mars.

suggests a timescale of 100,000 years. The main components of the terrestrial biosphere—water, carbon dioxide and nitrogen—are crucial to terraforming. If they do not occur in sufficient quantities, they would have to be transported from elsewhere.

The first step in terraforming would be to pour gases into the Martian atmosphere that would absorb more of the Sun's heat and eventually warm the whole planet. Would this be possible? Mankind's prodigious burning of fossil fuels over many centuries has already increased the levels of carbon dioxide within our atmosphere. This has led to the 'greenhouse effect', in which the atmosphere as a whole traps the Sun's heat more effectively, raising the average global temperature. It has been suggested that the greenhouse effect on Mars could be enhanced by pumping gases containing fluorine, chlorine and bromine into the atmosphere. The hope is that the average temperature of the Martian surface could be increased to the point where water vapour and carbon dioxide would be released from the polar caps and

from the ground. This would trap even more of the Sun's heat.

Once the greenhouse effect had been triggered, it would gather pace and eventually make Mars a warmer, wetter world than it is today, enveloped in a thick carbon dioxide atmosphere. This so-called 'manmade phase' would be relatively short, requiring perhaps a century or so.

The next task would be to introduce terrestrial organisms, perhaps artificially adapted to Martian conditions by genetic engineering. The carbon dioxide atmosphere could be converted to breathable oxygen by hardy plants such as mosses, algae, and lichens. The action of sunlight on the oxygen would probably create an ozone layer, which would absorb much of the harmful ultraviolet radiation from the Sun. Other dangerous radiation would eventually be absorbed by the sheer bulk of the atmosphere, as happens on Earth.

Though there is very little nitrogen on Mars, some nitrogen-fixing organisms have been found on Earth that could easily tolerate the levels found there.

The 'biological phase' would

take the 100,000 years mentioned above. By the end of this time open settlements will have spread across the globe and water will be able to flow freely on Mars.

Some scientists dismiss these ideas as nonsense. They assert that Martian gravity is too weak to hold onto a thicker atmosphere for any length of time, which is why the planet is the way it is today. Evidence suggests that Mars once had a thicker, wetter atmosphere, which seeped away into space, leaving only the thin blanket of carbon dioxide we see today. If nothing can be done to prevent such a process repeating itself, the colonists would be condemned to living in domes and underground dwellings.

Dreams of the future

Discussions of colonizing and terraforming, of antimatter propulsion and self-replicating robots, may seem wildly speculative. But only a generation ago the very thought of space travel and of men on the Moon would have been viewed similarly. The science-fiction writer Arthur C Clarke has noted that forecasts of future developments are invariably too optimistic in the short term but too pessimistic in the long. Clarke's own record in this field should not be overlooked: in 1945 he proposed the communications satellite, one of the vitally important products of the space age.

As long ago as 1952, in his first novel, *The Sands of Mars*, Clarke described a first base on the planet with uncanny prescience. Starved of funds from the Earth, the pioneers set about terraforming the planet to ensure its long-term independence.

Clarke's dream will be fulfilled during the 21st century: the exploration of Mars will inevitably lead to its colonization. Mars represents more than just another space research goal: it is the next evolutionary step of our species.

If terraforming proves successful, Mars may once again boast a warmer climate, allowing water to flow freely across its surface.

ACKNOWLEDGEMENTS

We wish to acknowledge the help of many scientists, engineers and other individuals around the world whose assistance was invaluable in the preparation and writing of this book.

ITN's original report on manned missions to Mars was compiled by Frank Miles and David Chater, and broadcast in February 1986. Gerry Webb, Managing Director of Commercial Space Technologies (CST) Ltd in London, made available analyses of Soviet space technology by Alan Bond and John Parfitt. After the maiden launch of Energia in May 1987, they revised their original analyses for us and freely gave of their time to help prepare material for this book. Dr John Niehoff of the Science Applications International Corporation (SAIC) office in Schaumburg, Illinois, provided unpublished material on VISIT orbits for ITN's original report, and was invaluable in updating that material for this book. Dr Alan Friedlander of SAIC, Charles Vick, Dr David Webb of the National Commission on Space and the Planetary Society also gave their time and assistance generously.

Nicholas Booth also wishes to acknowledge the untiring advice and assistance of Dr Al Hibbs, formerly of the Jet Propulsion laboratory, throughout the writing of this book. Fran Waranius and Stephen Tellier of the Lunar and Planetary Institute also provided much needed help. The editors wish to thank the following people for their support; they lived with this book as it took shape and helped amend the material. They are: Andy Bartram, Malcolm Beatson, Barbara Brooker, Susan Davis, Jill Gascoigne, Gail Hawker, Christina di Julio, John Mason, Anna Vadaketh and Alison Wood.

The editors are also indebted to the many individuals listed below, who provided invaluable information and advice:

PART ONE
Dr John Aaron, Office of Exploration, NASA Headquarters, Washington DC; Dr Buzz Aldrin, former Apollo astronaut and consultant to SAIC in Hermosa Beach, California; Don Bane, Public Affairs Office, Jet Propulsion Laboratory, Pasadena, California; Professor Jacques Blamont, scientific consultant to the Centre National d'Etudes Spatiales, Paris; Dr Roger Bourke, Project Manager, Mars Sample Return Mission, Jet Propulsion Laboratory; Dr John Butler, NASA Marshall Space Flight Center, Huntsville, Alabama; Dr Benton C Clark, Martin Marietta, Denver, Colorado; Dr Pamela Clark, Jet Propulsion Laboratory; Phil Clark; Dr Genevieve Debouzy, Senior Scientist, CNES, Paris; Dr Michael Duke, Chief, Earth and Planetary Sciences Division, NASA Johnson Space Center, Houston, Texas; Dr Larry Esposito, Laboratory for Atmospheric and Space Physics, University of Colorado, Boulder, Colorado; Dr Kyle Fairchild, Advanced Programs Office, NASA Johnson Space Center; Dr Jim French, American Rocket Company, Menlo Park, California; Dr Louis Friedman, Executive Director, The Planetary Society, Pasadena, California; Dr Tom Paine, Chairman, National Commission on Space; Dr Bill Purdie, Project Manager, Mars Observer, Jet Propulsion Laboratory; Dr Barney Roberts, Missions Manager, Advanced Programs Office, Johnson Space Center; Dr Thomas Thorpe, Science Manager, Mars Observer, Jet Propulsion Laboratory; Scott Webster, Orbital Sciences Corporation, Fairfax, Virginia.

PART TWO
Dr Per Anders Hansson, Life Sciences Director of Commercial Space Technologies, agreed to many lengthy interviews and provided much insight into the physiological problems associated with spaceflight. Squadron Leader Richard Harding of the RAF Institute of Aviation Medicine, Farnborough, and Dr Helen Ross of the Department of Psychology, University of Stirling, provided much advice and read through the original drafts. Anne-Maria Brennan was frequently on hand to provide invaluable briefing and background notes.

We also wish to thank:

Dr John Billingham, Life Sciences Directorate, NASA Ames Research Center, Mountain View, California; Dr B J Bluth, NASA Headquarters; Dr Yvonne Clearwater and Dr Mary Connors, NASA Ames Research Center; Dr Albert Harrison, University of California at Davis; and Dr Patricia Santy of NASA Johnson Space Center.

PART THREE
Dr David Hill and Dr John Zarnecki of the Unit for Space Sciences, Physics Department, University of Kent at Canterbury, for their computer modelling of the interplanetary dust problem; Dr Johnny Kwok, Jet Propulsion Laboratory; and Dana Rotegard of the University of Minnesota for information on Δv concepts.

PART FOUR
Dr Victor Baker, University of Arizona, Tucson, Arizona; Dr Michael Carr, US Geological Survey, Menlo Park, California; Dr Stephen Clifford, Lunar and Planetary Institute, Houston,

Texas; Dr Mary Dale-Bannister and Dr Edward Guinness of the Department of Earth and Planetary Science, University of Washington in St Louis, Missouri; Dr John Guest, University of London Observatory; Dr Harold Masursky, US Geological Survey, Flagstaff, Arizona; Dr James Tillman, University of Washington, Seattle.

PARTS FIVE AND SIX

Dr Bart Hibbs, AeroVironment Inc, Monrovia, California; Montye C Male of TRW Space and Defense, Redondo Beach, California; Dr Christopher McKay, Dr David Smernoff and Dr Carol Stoker, NASA Ames Research Center; Dr Don Morrison, NASA Johnson Space Center; Lori Stiles, Public Information Office, University of Arizona, Tucson.

Further reading

The following list is not intended as an exhaustive bibliography; rather it represents a few of the most useful books about Mars and its future exploration.

NASA REPORTS

For sale by the Superintendent of Documents, US Government Printing Office, Washington, DC 20402:

Planetary Exploration Through Year 2000: An Augmented Program (1986).

A colourful and non-technical guide to future solar system exploration, with a good section about Mars.

The Martian Landscape (1978), NASA SP–425.

A large-format, glossy book with most of the early Viking lander surface views. An introduction by Tim Mutch, leader of the Lander Camera team, provides an insight into the problems of taking pictures on Mars.

Atlas of Mars: The 1:5,000,000 Map Series (1979), NASA SP–438.

Another large-format book containing all the maps and technical data from Viking orbiter images, compiled by US Geological Survey cartographers.

Viking Orbiter Views of Mars (1980), NASA SP–441.

Another picture-book of Mars, with fairly technical information about the individual orbiter frames.

Manned Mars Missions: Working Group Papers (two volumes), NASA M002 June 1986.

A highly technical review of every aspect of Mars exploration, following a year-long study between NASA and the Los Alamos National Laboratory.

GENERAL BOOKS

Penelope Boston (ed.), *The Case For Mars* (1984), American Astronautical Society, PO Box 28130, San Diego, California 92128.

A technical review of the first Case for Mars conference held at the University of Colorado in 1981. Less exhaustive than NASA's 1986 study but contains many important ideas, further developed at the second conference (see below).

Michael H Carr, *The Surface of Mars* (1981), Yale University Press, New Haven and London.

The definitive guide to Mars, written by the leader of the Viking orbiter TV experiment. Though technical, it is large-format and exhaustive in its treatment of every aspect of the red planet.

Lunar Bases and Space Activities of the 21st Century (1985), Lunar and Planetary Institute, 3303 NASA Road One, Houston, Texas 77058.

A technical review of a 1985 conference, with a good section on aspects of a Mars base.

Christopher McKay (ed.) *The Case For Mars II* (1985), AAS, San Diego, California.

Larger than its predecessor, this is the technical review of the 1984 Case for Mars conference.

James Oberg, *Mission to Mars* (1982), New American Library, New York.

This book was based on the proceedings of the first Case for Mars conference, and is much less technical than the AAS publication above. The author is a noted space writer who works at the Johnson Space Center in Houston.

Pioneering The Space Frontier (1986), Bantam Books, New York.

The report of the US National Commission on Space. A glossy, almost coffee-table book, it is a good guide to future human expansion into the solar system.

Mark Washburn, *Mars at Last!* (1977), Abacus Books, London.

An easy-to-read guide to the exploration of Mars from the first observations by the Babylonians to the excitement of the early part of the Viking mission.

ITN will be covering all developments in space, particularly those related to Mars, in its bulletins. *Aviation Week and Space Technology* in the US and *Spaceflight News* in Britain provide more technical information on spacecraft; *Sky and Telescope* in the US and *Astronomy Now* in Britain will cover the planet and its moons.

The Planetary Society is very active in supporting Mars exploration. Its bimonthly *Planetary Report* covers many aspects of future space exploration, and the Society distributes *The Mars Underground News*. For more information, write to:

The Planetary Society
65 North Catalina Avenue
Pasadena CA 91106.

189

Picture credits

The Soviet pictures contained in this book were provided by Novosti in London; we are grateful to Blanche La Guma for her assistance in locating many of the more unusual pictures.

The NASA material was kindly provided by many of its centres across the US: the expertise of its public information officers in finding rare material is legendary. We gratefully acknowledge the assistance of the following individuals: Jurrie van der Woude of the Jet Propulsion Laboratory; Mike Gentry, Cindy Leos and Lisa Vasquez of the Johnson Space Center; and Marie Jones and Althea Washington of NASA Headquarters.

The maps reproduced here were kindly provided by the US Geological Survey in Flagstaff, Arizona; Ray Batson and Haig Morgan provided information regarding their availability. Kathy Teague of the Planetary Data Facility went to much trouble in locating them, as well as many of the high-resolution Viking orbiter views reproduced in Part Four. Debra Rueb at the Lunar and Planetary Institute also helped locate more unusual material. We are also grateful to Dr Mary Dale-Bannister of the University of Washington in Missouri for allowing us to use many of her high-resolution Viking lander images.

Finally, we are grateful to Carter Emmart and Mike Carroll for allowing us to use their artwork material and to Chris Lyons for his original artwork specially commissioned for this book.

The detailed list of illustrations include the following abbreviations:
NASA HQ—NASA Headquarters, Washington, DC;
NASA JPL—NASA Jet Propulsion Laboratory, Pasadena, California;
NASA JSC—NASA Johnson Space Center, Houston, Texas;
USGS—US Geological Survey, Flagstaff, Arizona;
Mary Dale-Bannister is in the Earth and Planetary Sciences Department of the University of Washington, St. Louis, Missouri:
John E Guest, a member of the Viking Orbiter TV Experiment Team, is at the University of London Observatory, Mill Hill, London.

Page 7: NASA JPL; 8: NASA JPL; 11: Novosti; 12: Main Picture—G E Perry, MBE Others—Novosti; 13: Novosti; 15: NASA HQ; 16: Upper—NASA JPL Lower—NASA HQ 17: NASA JPL; 19: Andrew Wilson; 30: NASA JSC; 31: NASA JSC; 32: Novosti; 33: NASA JSC; 34: Orbital Sciences Corporation; 35: Orbital Sciences Corporation; 36: RCA; 37: NASA JSC; 43: Novosti; 53: NASA JSC; 54: NASA JSC; 55: Novosti; 56: NASA JSC; 57: NASA JSC; 58: Top—NASA JSC Bottom—Novosti; 60: Novosti; 61: NASA HQ; 62: Top Left—Novosti Top Right—Andy Wilson Bottom—Novosti 63: Top—NASA JSC Middle—Andy Wilson Bottom: Novosti 65: NASA HQ; 67: NASA JSC; 68: Novosti; 69: NASA JPL; 71: Novosti; 72: NASA JSC; 73: Carter Emmart; 74: NASA JSC; 75: Top—NASA JPL Bottom—NASA JSC; 76: NASA JSC; 77: Left—Novosti Right—NASA JSC; 80: Carter Emmart; 82: NASA JPL; 83: NASA JPL; 84: NASA HQ; 86: NASA HQ; 88: Carter Emmart; 90: Left—NASA JPL Right—Mary Dale-Bannister; 91: Mary Dale-Bannister; 92: NASA JPL; 93: NASA JPL; 94: NASA JSC; 95: Top—NASA HQ Bottom—NASA JSC; 96: NASA JSC; 97: Left and Inset—NASA JSC Right—Andy Wilson; 99: NASA/Science Photo Library; 101: USGS; 102: NASA/Science Photo Library; 103: NASA JPL; p. 104—107: All maps provided by USGS; 108: NASA/Science Photo Library; 110: USGS; 111: Lower—USGS All others—JPL; 112: NASA JPL; 113: Upper—NASA/Science Photo Library Lower—NASA JPL; 114: NASA JPL; 115: NASA JPL; 116: Lower—Lunar and Planetary Institute; Middle and inset—NASA JPL; 117: NASA JPL; 118: NASA JPL; 119: Mary Dale-Bannister; 120: Top—John E. Guest; Bottom: NASA JPL; 121: Top—Lunar and Planetary Institute; Bottom – John E. Guest; 122: NASA JPL; 123: USGS; p. 124—126: John E. Guest; 127: NASA JPL; 128: NASA JPL; 129: Main—John E. Guest Inset—Lunar and Planetary Institute; 130: NASA JPL; 131: USGS; 132: Top—USGS Bottom – NASA JPL; 133: NASA JPL; 134: NASA JPL; 135: NASA JPL; 136: NASA/Science Photo Library; 137: Top— USGS Bottom—NASA JPL; 138: John E. Guest; 139: Top—John E. Guest Bottom—NASA JPL; 140: NASA JPL; 141: NASA JPL; 142: NASA JPL; 143: NASA JPL; 145: Mary Dale-Bannister; 146: NASA JSC; 148: USGS; 149: Left—Mary Dale-Bannister; Right and Lower Left—NASA JSC; 150: Top—NASA JPL; Bottom—NASA/Science Photo Library; 151: Top—TRW Inc. Bottom—NASA JPL; 152: Mary Dale-Bannister; 153: Top Left and Right—NASA JPL; Bottom Left—Mary Dale-Bannister; Bottom Right—NASA/ Science Photo Library; 154: NASA/ Science Photo Library; 155: Michael Carroll; 156: NASA HQ; 157: Lori Stiles, University of Arizona; 158 and 159: Michael Carroll; 162: NASA HQ; 163: NASA JSC; 164: NASA JSC; 165: NASA JPL; 166 and 167: Michael Carroll; 169: NASA HQ; 170: NASA HQ; 171: Top—NASA HQ; Bottom—NASA JSC; 173: NASA JSC; 174: Boeing Aerospace; 177: Michael Carroll; 178—180: Carter Emmart; 181: Top—Carter Emmart Bottom—Novosti; 182: NASA JPL; 183—185: NASA JSC; 186: NASA JPL; 187: Michael Carroll.

INDEX